Lecture Notes in Electrical Engineering

Volume 510

**** Indexing: The books of this series are submitted to ISI Proceedings, EI-Compendex, SCOPUS, MetaPress, Springerlink ****

Lecture Notes in Electrical Engineering (LNEE) is a book series which reports the latest research and developments in Electrical Engineering, namely:

- Communication, Networks, and Information Theory
- Computer Engineering
- Signal, Image, Speech and Information Processing
- Circuits and Systems
- Bioengineering
- Engineering

The audience for the books in LNEE consists of advanced level students, researchers, and industry professionals working at the forefront of their fields. Much like Springer's other Lecture Notes series, LNEE will be distributed through Springer's print and electronic publishing channels.

For general information about this series, comments or suggestions, please use the contact address under "service for this series".

To submit a proposal or request further information, please contact the appropriate Springer Publishing Editors:

More information about this series at http://www.springer.com/series/7818

Maxine Eskenazi · Laurence Devillers
Joseph Mariani
Editors

Advanced Social Interaction with Agents

8th International Workshop on Spoken Dialog Systems

Editors
Maxine Eskenazi
Language Technologies Institute
Carnegie Mellon University
Pittsburgh, PA, USA

Joseph Mariani
LIMSI-CNRS
Paris-Saclay University
Orsay, France

Laurence Devillers
LIMSI-CNRS
Sorbonne University
Paris, France

ISSN 1876-1100 ISSN 1876-1119 (electronic)
Lecture Notes in Electrical Engineering
ISBN 978-3-030-06364-1 ISBN 978-3-319-92108-2 (eBook)
https://doi.org/10.1007/978-3-319-92108-2

This Springer imprint is published by the registered company Springer Nature Switzerland AG
The registered company address is: Gewerbestrasse 11, 6330 Cham, Switzerland

Preface

The International Workshop on Spoken Dialogue Systems (IWSDS) series provides an international forum for the presentation of research and applications as well as a place where lively discussions can take place between academic and industrial researchers. IWSDS focusses on the practical implementation of spoken dialogue systems in everyday applications.

Following the success of IWSDS'09 (Isree, Germany), IWSDS'10 (Gotemba Kogen Resort, Japan), IWSDS'11 (Granada, Spain), IWSDS'12 (Ermenonville, France), IWSDS'14 (Napa, USA), IWSDS'15 (Busan, South Korea) and IWSDS'16 (Saariselka, Finland), the Eighth IWSDS, IWSDS'17 took place in Farmington PA (USA), from 6 to 9 June 2017.

This book consists of the revised versions of a selection of the papers that were presented at IWSDS'17.

As intelligent agents gain greater ability to carry on a dialogue with users, several qualities emerge as being essential parts of a successful system. Users do not carry on long conversations on only one topic—they tend to switch amongst several topics. Thus, we see the emergence of multi-domain systems that enable users to seamlessly hop from one domain to another. The systems have become active social partners. Thus, work on social dialogue has become crucial to active and engaging human–robot/agent interaction. As these new systems come into being, they need some coherent framework that guides their structure as chatbots and conversational agents. That structure will naturally lend itself to being the human–robot/agent assessment mechanism. As these systems increasingly assist humans in a multitude of tasks, the ethics of their existence, their design and their interaction with users becomes a crucial issue.

We gave this year's workshop the theme, "*Advanced Social Interaction with Agents*", which covers:

- Chatbots and Conversational Agents
- Multi-domain Dialogue Systems
- Human-Robot Interaction
- Social Dialogue Policy
- Advanced Dialogue System Architecture

A special session on *Chatbots and Conversational Agents* was organized within the conference, as part of a series of workshops and special sessions on that topic, which was initialized at the NII-Shonan seminar on "*The Future of Human–Robot Spoken Dialogue: From Information Services to Virtual Assistants*" in 2015, which already included a first Workshop on "*Collecting and Generating Resources for Chatbots and Conversational Agents—Development and Evaluation*" (RE-Wochat) in the framework of the "*Language Resources and Evaluation Conference*" (LREC'2016) conference in Portorož (Slovenia, May 2016). A second Workshop on "*Chatbots and Conversational Agents*" (Wochat) took place in Los Angeles CA, in September 2016, within the "*Intelligent Virtual Agents*" conference (IVA'2016) conference. The third meeting was at IWSDS'17 where seven papers were selected, including two position papers.

Based on research in psychology that suggests that people tend to seek companionship with those who have similar levels of extraversion, Jacqueline Brixey and David Novick in their Chapter "Building Rapport with Extraverted and Introverted Agents" explore whether it is also the case when considering the interaction between humans and Embodied Conversational Agents (ECA). For this, they created ECAs representing an extravert and an introvert and conducted experiments in an immersive environment with subjects demonstrating various levels of extraversion. They conclude that subjects did not report the greatest rapport when interacting with the agent most similar to their level of extraversion.

In their chapter "Automatic Evaluation of Chat-Oriented Dialogue Systems Using Large-Scale Multi-references", Hiroaki Sugiyama, Toyomi Meguro and Ryuichiro Higashinaka, from NTT, address the open problem of the automatic evaluation of chat-oriented dialogue systems. They propose a regression-based automatic evaluation method that evaluates the utterances generated by the system, based on their similarity to many reference sentences and their annotated evaluation values. The results they obtain with their approach show a higher sentence-wise correlation with human judgment than previous methods such as Discriminative BLEU.

Koh Mitsuda, Ryuichiro Higashinaka and Yoshihiro Matsuo, also from NTT, authored the chapter "What Information Should a Dialogue System Understand?: Collection and Analysis of Perceived Information in Chat-Oriented Dialogue". For this, they collected and clustered information that humans perceive from each utterance in chat-oriented dialogues. They categorized the types of perceived information according to four hierarchical levels and evaluated the inter-annotator agreement, which appeared to be substantial. This result confirmed their opinion that their categorization is valid and allowed them to clarify the types of information that a chat-oriented dialogue system should understand.

Emer Gilmartin, Benjamin R. Cowan, Carl Vogel and Nick Campbell in their chapter "Chunks in Multiparty Conversation—Building Blocks for Extended Social Talk" consider the various phases in long (c. one hour) multiparty casual conversations, making the difference between "chat" phases, which are highly interactive, and "chunk" phases, where one speaker takes the floor, or when the conversation moves through predictable stages. The authors describe the collection of long-form conversations and provide preliminary results showing that chat and chunk segments exhibit differences in the distribution of their duration and that chat varies more while chunks have a stronger central tendency.

Geetanjali Rakshit, Kevin K. Bowden, Lena Reed, Amita Misra and Marilyn Walker indicate that chatbots are a rapidly expanding application of dialogue systems, for example, bot services for customer support, or for users interested in casual conversations with artificial agents. Based on the assumption that many people love nothing more than a good argument, they used a corpus of argumentative dialogues on three topics (gay marriage, gun control and death penalty) annotated for agreement and disagreement, stance, sarcasm and argument quality. They then explored the use of their arguing bot prototype, in "Debbie, the Debate Bot of the Future", which selects arguments from conversational corpora and attempts to use them appropriately in context.

In addition to these presentations, a panel discussion moderated by Joseph Mariani (LIMSI-CNRS) included Kevin Bowden (UCSC), Laurence Devillers (LIMSI-CNRS), Satoshi Nakamura (NAIST), Alexander Rudnicky (CMU), Marilyn Walker (UCSC) and Zhou Yu (CMU). The panel addressed nine questions on three issues (**chatbot design, human users involvement and evaluation**), which received feedback from the panelists and the audience during the panel discussion and beforehand in two position papers.

In their position paper, "Data-Driven Dialogue Systems for Social Agents", Kevin K. Bowden, Shereen Oraby, Amita Misra, Jiaqu Wu, Stephanie Lukin and Marilyn Walker argue that in order to build dialogue systems to tackle the ambitious task of holding social conversations, we need a data-driven approach that includes insights into human conversational "chit-chat" and that incorporates different NLP modules. Their strategy is to analyze and index large corpora of social media data and couple this data retrieval with modules that perform tasks such as sentiment and style analysis, topic modelling and summarization, with the aim of producing more personable social bots.

Alex I. Rudnicky proposes to use the term "ChaD: Chat-Oriented Dialog Systems" to refer to the recently developed non-goal-directed dialogue systems that differ from the ones that proceed them, which only performed goal-directed interaction. He defines CHAD systems as having a focus on social conversation where the goal is to interact while maintaining an appropriate level of engagement with a human interlocutor and raises the need to address several research challenges: continuity, dialogue strategies and evaluation.

Following is a brief report on the panel discussion:

Chatbot Design

Q1. In your opinion, what is the most important innovation needed to improve chat dialogue, and why?
Q2. How good are existing computational tools to support chatbot development work? What are their main advantages and limitations?

Satoshi Nakamura advocates the use of a layered approach to design dialogue systems, including a physical layer, an utterance layer, a dialogue layer (with language understanding and generation, and semantics), a discourse layer and an intention layer. The most important innovations needed are theory (which would decrease the need for huge amounts of data) and evaluation measures at each level, together with annotated corpora of spoken and text dialogues and discourse, ontology, and knowledge, information sharing between chat and task dialogues, as well as emotion and symbol and mutual grounding information. Those groundings would be shaped by never-ending learning.

For Alex I. Rudnicky, just as humans have conventions for communication management, systems should follow human conventions. Conversation is a mixture of dialogue types, supporting different needs and goals: tasks, social dialogue, instrumental dialogue such as persuasion. This requires a new conversation management for non-task-based dialogue, taking into account an overall conversation structure, turn-to-turn shifts and goal blending within turns. He expressed the opinion that it is important *to have a better understanding of how different levels of chat dialogue operate in concert: humans are quite good at interleaving levels, but automatic systems are not. At the same time there, is no good theory of the interlocutor and the mechanisms that allows them to manage different levels of interaction in service of the goals that they support. It is tempting to treat different levels separately, and this approach was quite useful in developing our understanding of goal-directed dialogue. But to approach the sophistication of human language-based interaction, we need to develop an integrated framework.*

Marilyn Walker expressed the opinion that the most needed innovation is the capacity to retain the memory of the user and models of past conversations, keeping an ongoing context of the conversation. This would allow for shared experience (between user and system) and increased chat dialogue naturalness. She also supports a better understanding of the structure of non-task-oriented dialogue: What dialogue strategies are relevant? What are the useful semantic relationships between dialogue turns? There is also a need for a deeper understanding of user utterances, a semantic mapping into a topical space.

For Laurence Devillers, it is necessary to understand the architecture of casual conversation, to develop descriptors of the above for annotation and learning and to find a way to maintain the conversation.

Q3. What is the "role of theories of dialogue" vs. "role of lots and lots of data"? Is deep learning going to save us all?

Zhou Yu does not believe that deep learning will save us all: *deep learning methods today are just like SVM back 10 years ago. They are just methods to help us to understand the world and create new tools and applications. Most importantly, deep learning currently only works pretty well with supervised tasks, which requires millions of well-labelled training data and performs poorly with limited labelled data. Conversations require human participation; therefore, it is really hard to collect such a big amount of labelled data.* She advocates a shift towards more transfer, unsupervised and semi-supervised learning and hopes that the community cares more about how to solve the problems, instead of focusing on shifting the problems to deep learning models.

Q4. Is chat dialogue a single genre, or can it be divided into different types (e.g. argument vs rapport building)?

Some of the panelists think that there is no clear genre distinction in human–human conversation: people are not conscious of their choice of register; information is carried over between the social and task parts, and the latter should help or affect each other. Others believe that genre can be distinguished by the level of formality, even if mixed, within the same dialogue.

The opinion of Marilyn Walker is that we have no good evaluation metrics because we do not have a good classification of genres, as these reflect different dialogue purposes: *chat dialogue is not a single genre, and much more progress would be made by acknowledging that. If we have subtypes of chat, then we could much more easily evaluate whether the chat system was achieving its goals. For example, we could have a chatbot that argues with people about social and political topics and measures how much more people know about different sides of an argument after the conversation. Or we could have a "celebrity gossip" chatbot that could then be evaluated by whether it provides the user with gossip they are interested in. Also, the idea of an "ongoing" conversant who has a memory for past conversations and what you said would almost result in a different genre. Even for domains such as film/movies that have a clearer structure, being able to move a conversation from movies to actors to actor spouses to gossip about the actors is beyond what we can currently easily do.*

Zhou Yu does not think that there is a single rule to define genre: *most human–human conversations are mixed with different genres at once. For example, when you go to the coffee room, you would start the conversation with some social talk, such as how are you doing? How's your weekend? Then, you slowly go to your work-related content. Also, the ranges of these conversations are not merely according to social convention, but also out of the consideration of effectiveness. For example, if your co-worker told you that her kid was sick over the weekend, then you'd better not ask him to stay late to finish some extra work.*

Human User Involvement

Q5. How can user expectations be properly set for a given chatbot platform? How important is user experience design with respect to the limitations of natural language understanding and dialogue management for a chat-oriented dialogue to be usable?

According to Satoshi Nakamura, the personae of system and user (age, family, language and history), and the sharing of this knowledge, are crucial.

For Kevin Bowden, what is missing from chatbots is the ability to make the dialogue more informal, adding hedge words. *It's not the best strategy for an agent to try to be human when they can't. Embrace the fact that you're a bot, and set appropriate expectations. Bots that are too polite are not that interesting. The goals of a chatbot are to engage experience and to communicate a message.*

Laurence Devillers stresses the importance of conducting a lot of experiments with a human in the loop: *humans use emotions. Machines need to have emotions, sociability, friendliness, cultural awareness.* Chatbots need culture, personality, emotion; but that can be confusing to people. The need to adapt to different users brings a danger: it could be used to exploit, abuse or influence individual people. There is an IEEE Working Group about manipulation and nudging. She also thinks that there are currently asymmetries between emotion generation and detection: there is a need for better coupling between those mechanisms.

Zhou Yu thinks that ideally every user should have a personal (personalized?) assistant. These can be bootstrapped based on demographic characteristics (age, culture). Privacy is an issue, but adaptability is still needed.

Marilyn Walker stresses the fact that designing multi-domain systems is a challenge, while multi-domain is almost a prerequisite for open-domain chatbots. David Novick thinks that conversations have a broader structure, almost like a story or narrative. How close are we to modelling this? M. Walker replies that we need a taxonomy of conversational moves: study socially aware people, how they pass the ball and keep a conversation going.

Emer Gilmartin reminds us that personal assistants perform several distinct human roles. How do humans relate to that? For David Traum, predictable does not equal usable: there is a need to get accurate expectations from the user. To which Laurence Devillers answers that *the user expectation is difficult to set properly in a chatbot platform, it means that the system should be adaptable to the habits of the user knowing the user's likes and dislikes for example. The user's experience design is important because the system is still not able to understand open-domain natural language and common sense as well as dialogue management should be more usable if the dialogue is adapted to the way the user speaks.*

Q6. What are the most important aspects of (human) chat dialogue that should be incorporated in chat dialogue systems?

For Kevin Bowden, important human aspects to consider include being less formal, with more colloquial speech, and exploring agents that are less polite.

For Laurence Devillers, *the most important aspect of human dialogue is to cope with affect/empathy but also irony/sarcasm/humour in order to build a more social, friendly and "chatty" chatbot. Drives and emotions such as excitement and depression are used to coordinate action throughout life. One other really important thing is to consider the cultural aspect of the language. Societies and individuals around the world have different ways to express intentions through conversation and paralinguistic cues (mimics, gestures), interpret silence, etc. These particularities could be incorporated into the social chatbot in order to transmit the intended message. So in order to build this chat dialogue, we have to give the chatbot a personality, a cultural sensibility and emotions.*

Q7. What kinds of techniques should be added (beyond simple corpus response retrieval- or rule-based template filling) to the chat system repertoire to create more human-like (or just more coherent, better) chat dialogues?

Kevin Bowden thinks that retrieval- and rule-based systems are great; they just need to have more fluid transient states to connect them, while for Laurence Devillers, *hybrid systems combining deep learning and rule-based templates will be more chatty and more coherent. New data sources could be used such as movies, everyday conversations in talk shows, Internet for training for social response generation. Also, symmetrical dialogue with a better coupling between natural language understanding and natural language generation is necessary.*

While, for Marilyn Walker, there are many open problems in multi-domain task-oriented dialogue, which if they were solved would be highly applicable to chat dialogue. *We need, for example, to understand how to represent a discourse agenda and integrate information across different applications. For example, in task-oriented dialogues, we do not know how to interleave search and QA with a directed dialogue such as:*

- *User: set a time*
- *System: for how long*
- *User: how long does it take to boil an egg* (goes to QA/search)
- *System: Hard boiled or soft boiled* (comes back from QA)
- *User: hard boiled*
- *System: I am setting the timer for 5 minutes for a hard-boiled egg.*

Q8. Some argue that task-oriented dialogue systems need not (and perhaps should not) be human-like, but should strive for efficiency. When it comes to chat dialogue systems, should similar arguments be made, or must they be more human-like to be successful?

Joseph Mariani asks about the *uncanny valley effect*: if the machine behaviour is too similar to humans, it can become annoying, even if it is agreed that the goal is not to build artefacts that are indistinguishable from humans, but systems that can interact with humans in a natural way.

Kevin Bowden thinks that when humans talk to a bot which is supposed to be human, they subconsciously recall learned expectations as to what a human can do, such as process emotion correctly: *since we currently cannot satisfy these expectations, it can hurt the user experience. I would suggest that our bots should embrace the fact they are a bot, and not try to present themselves as human, to perhaps lower expectations, subsequently increasing the overall experience. In fact, I think it could be quite interesting to try other means of representation. However, there is of course a trade-off, if you make your bot less human, you need to insure that there is some other functionality which engenders engagement.*

The opinion of Zhou Yu is that good conversational systems are not evaluated by how human-like they are, but how easily they elicit natural human (user) behaviour. *If we become able to reduce the cognitive load of the user, more and more people will be willing to use our systems.*

Evaluation

Q9. What kinds of methods can be used to efficiently evaluate human, human–computer and computer–computer chat dialogue?

For Alex Rudnicky, objective, automatic measures are necessary for learning: *we do not have a theory of what a good chat is, either turn-by-turn or as a global metrics. We need to model continuity, engagement.* Zhou Yu agrees that the longer a person is willing to talk to a system, the better. But this measure is not automatic and not reproducible.

According to Laurence Devillers, we need to compare interactions between humans, between machines and between humans and machines. But there is no clear measure of what constitutes a good dialogue. Maybe using the Turing test to evaluate human interaction with the machine?

David Novick proposes that we learn from video game evaluation: *there is a contrast between game and task evaluation—games want to maximize engagement, not minimize it.* Also for Laurence Devillers, methods based on games like *jeux de roles* (role-playing games) could be used to efficiently evaluate human, human–computer and computer–computer chat dialogue, while *we also need to create new kinds of measures in the dialogue like social engagement in interaction, symmetry between agent and user, degree of familiarity or intimacy.*

For Alex Rudnicky, a key missing element in social chat research is a clear understanding of measurement. *For goal-directed systems, it was much easier as task completion, achieving a goal, was an objective criterion that could be agreed on. Additional criteria, such as time-to-completion and number-of-turns, were also accepted in part because they could be measured. Non-goal-oriented chat is different: sometimes the goal is to operate within an expected social structure ("hello", "goodbye", "thank you", etc.), or to develop a relationship (by establishing a common ground and history), or simply a way to pass the time (while practicing conversational skill). The challenge is to find objectively measurable criteria for these and to move beyond judgment-based evaluation. Doing so will create the necessary foundation for learning-based approaches.*

For his part, Satoshi Nakamura advocates an evaluation via brain activity measurement to test the naturalness of dialogues as perceived by humans.

A special session on *Multi-domain Dialogue Systems* took place with the goal of examining the state of the art in the newly emerging area of multi-domain systems. These systems can be characterized as combining more than one slot-filling or chatbot system. Past dialogue systems have concentrated on one domain, making the task of deciding what the user is asking for (filling a slot, understanding an intention, determining the question in a Q/A set-up) much easier. These systems have one central decision mechanism that interfaces with the user, determines the domain of the user's request (or if it is out of domain) and then sends the user utterance on to the dialogue system that deals with the domain in question (or with all out-of-domain requests). The domain decision mechanism can be a rule-based module, the easiest to set up for a new system, or if there is sufficient data, it can use neural nets to represent the request/action pair learned from past dialogues. The latter approach was extremely rare as of the writing of this introduction as the domain is new and there is little available data. Due to the very large amount of data needed to train such a system, and the need to cover many domains if the data is to be quasi-universally useful, this dataset is not easy to fashion.

There were three chapters in this session. The chapter by Alexandros Papangelis, Stefan Ultes and Yannis Stylianou, "Domain Complexity and Policy Learning in Task-Oriented Dialogue Systems", deals with multi-domain systems comprising uniquely slot-filling systems. It compares two well-known policy learning algorithms, GP-SARSA and DQN in order to understand something about the nature of the varied domains that we try to group in a multi-domain system. What feature or features of a domain should we pay the most attention to when using complex statistics to model domain selection? They find that a requestable slot's entropy, the number of requestable slots in a given domain and database coverage are the main factors that play a role in whether a given domain will be referred to correctly and at the right time in a multi-domain system.

The chapter by Alexandros Papangelis and Yannis Stylianou, "Single-Model Multi-domain Dialogue Management with Deep Learning", asks how to create the mechanism that decides which domain the user wants. They show that instead of training the system using deep Q networks on one model for each domain in the system so far (multi-domain models), it is far better to train the system with one single domain-independent policy network. In this way, new domains can be added much more easily.

The chapter by Kyusong Lee, Tiancheng Zhao, Stefan Ultes, Lina Rojas-Barahona, Eli Pincus, David Traum and Maxine Eskenazi, "An Assessment Framework for DialPort", discusses how to assess a multi-domain system and uses the DialPort portal as its example. Assessment can be approached from both the point of view of the system developer and of the user. This chapter proposes a comprehensive assessment framework that encompasses both points of view. The framework includes response delay, smooth transition from one agent (system) to another and the ability to correctly choose the system that the user is to be connected to.

There were four chapters presented in the special session on *Human–Robot Interaction*. All of the chapters deal with the use of dialogue frameworks with robots.

The chapter by Lucile Bechade, Guillaume Dubuisson-Duplessis, Gabrielle Pittaro, Mélanie Garcia and Laurence Devillers, "Towards Metrics of Evaluation of Pepper Robot as a Social Companion for the Elderly", explains that field studies in human–robot interaction are necessary for the design of socially acceptable robots. Constructing dialogue benchmarks can have a meaning only if researchers take into account the evaluation of the robot, the human and of their interaction. This chapter describes a study aimed at finding an objective evaluation procedure for the dialogue with a social robot. The goal is to build an empathic robot (JOKER project), and it focuses on elderly people, the end-users expected by the ROMEO2 project. The authors carried out three experimental sessions. The first time, the robot was NAO, and it was with a Wizard of Oz (emotions were entered manually by experimenters as input to the program). The other times, the robot was Pepper, and it was totally autonomous (automatic detection of emotions and consequent decision). Each interaction involved various scenarios dealing with emotion recognition, humour, negotiation and cultural quiz.

The chapter by Juliana Miehle, Ilker Bagci, Wolfgang Minker and Stefan Ultes, "A Social Companion and Conversational Partner for the Elderly", presents the development and evaluation of a social companion and conversational partner for the specific user group of elderly persons. With the aim of designing a user-adaptive system, the authors create a system that responds to the desires of the elderly which have been identified during various interviews. They create a companion that talks and listens to elderly users.

The chapter by Onno Kampmann, Farhad Bin Siddique, Yang Yang and Pascale Fung, "Adapting a Virtual Agent to User Personality", proposes to adapt a virtual agent called "Zara the Supergirl" to the user's personality. User personality is detected using two models, one based on raw audio and the other based on transcribed text. Both models show good performance, with an average F-score of 69.6 for personality perception from audio, and an average F-score of 71.0 for recognition from text. Both models deploy a convolutional neural network. The study suggests that the openness user personality trait correlates especially well with a preference for agents with a more gentle personality. People also sense more empathy and enjoy better conversations when agents adapt to their personality.

The chapter by Pierrick Milhorat, Divesh Lala, Koji Inoue, Tianyu Zhao, Masanari Ishida, Katsuya Takanashi, Shizuka Nakamura and Tatsuya Kawahara, "A Conversational Dialogue Manager for the Humanoid Robot ERICA", presents a dialogue system for a conversational robot, Erica. The goal is for Erica to engage in more human-like conversation, rather than being a simple question-answering robot. The dialogue manager integrates question-answering with a statement response component which generates dialogue by asking about focus words detected in the user's utterance, and a proactive initiator which generates dialogues based on events detected by Erica. The authors evaluate the statement response component and find that it produces coherent responses to a majority of user utterances taken from a human–machine dialogue corpus. An initial study with real users also shows that it reduces the number of fallback utterances by half. This system is beneficial for producing mixed-initiative conversations.

The session on *Social Dialogue Policy* had seven chapters presented.

The chapter by Nurul Lubis, Sakriani Sakti, Koichiro Yoshino and Satoshi Nakamura, "Eliciting Positive Emotional Impact in Dialogue Response Selection", shows how introducing emotion in human–computer interaction (HCI) has created various system's abilities that can benefit the user. Among these is emotion elicitation, which has great potential to provide emotional support. To date, work on emotion elicitation has only focused on the intention of elicitation itself, e.g. through emotion targets or personalities. In this chapter, the authors extend the existing studies by using examples of human appraisal in spoken dialogue to elicit a positive emotional impact. They augment the widely used example-based approach with emotional constraints: (1) emotion similarity between user query and examples and (2) potential emotional impact of the candidate responses. Text-based human subjective evaluation with crowdsourcing shows that the proposed dialogue system elicits overall a more positive emotional impact and yields higher coherence as well as emotional connection.

The chapter by Svetlana Stoyanchev, Soumi Maiti, and Srinivas Bangalore, "Predicting Interaction Quality in Customer Service Dialogs", shows how to apply a dialogue evaluation *Interaction Quality (IQ)* framework to human–computer customer service dialogues. The IQ framework can be used to predict user satisfaction at the utterance level in a dialogue. Such a rating framework is useful for online adaptation of dialogue system behaviour and increasing user engagement through personalization.

The chapter by Bethany Lycan and Ron Artstein, "Direct and Mediated Interaction with a Holocaust Survivor", presents new dimensions in a testimony dialogue system that was placed in two museums under two distinct conditions: docent-led group interaction and free interaction with visitors. Analysis of the resulting conversations shows that docent-led interactions have a lower vocabulary and a higher proportion of user utterances that directly relate to the system's subject matter, while free interaction is more personal in nature. Under docent-led interaction, the system gives a higher proportion of direct appropriate responses, but overall correct system behaviour is about the same in both conditions because the free interaction condition has more instances where correct system behaviour is to avoid a direct response.

The chapter by Zahra Rahimi, Diane Litman and Susannah Paletz, "Acoustic-Prosodic Entrainment in Multi-party Spoken Dialogues: Does Simple Averaging Extend Existing Pair Measures Properly?", explains how linguistic entrainment, the tendency of interlocutors' speech to become similar to one another during spoken interaction, is an important characteristic of human speech. Implementing linguistic entrainment in spoken dialogue systems helps to improve the naturalness of the conversation, likability of the agents, and dialogue and task success. The first step toward the implementation of such systems is to design proper measures to quantify entrainment. Multi-party entrainment and multi-party spoken dialogue systems have received less attention compared to dyads.

The chapter by Patrick Ehrenbrink and Stefan Hillmann, "How to Find the Right Voice Commands? Semantic Priming in Task Representations", discusses a common way of interacting with conversational agents: through voice commands. Finding appropriate voice commands can be a challenge for developers and often requires empirical methods where participants come up with voice commands for specific tasks. Textual and visual task representations were compared in an online study to assess the influence of semantic priming on spontaneous, user-generated voice commands.

The chapter by Louisa Pragst, Wolfgang Minker and Stefan Ultes, "Exploring the Applicability of Elaborateness and Indirectness in Dialogue Management", investigates the applicability of soft changes to system behaviour, namely changing the amount of elaborateness and indirectness. To this end, the authors examine the impact of elaborateness and indirectness on the perception of human–computer communication in a user study. Here, they show that elaborateness and indirectness influence the user's impression of a dialogue and discuss the implications of their results for adaptive dialogue management. The authors conclude that elaborateness and indirectness offer valuable possibilities for adaptation and should be incorporated in adaptive dialogue management.

The chapter by Zhou Yu, Vikram Ramanarayanan, Patrick Lange and David Suendermann-Oeft, "An Open-Source Dialog System with Real-Time Engagement Tracking for Job Interview Training Applications", explains how, in complex conversation tasks, people react to their interlocutor's state, such as through uncertainty and engagement, to improve conversation effectiveness. If a conversational system reacts to a user's state, would that lead to a better conversation experience? To test this hypothesis, the authors designed and implemented a dialogue system that tracks and reacts to a user's state, such as engagement, in real time. Experiments suggest that users speak more while interacting with the engagement-coordinated version of the system as compared to a non-coordinated version. Users also reported the former system as being more engaging and providing a better user experience.

There are four chapters in the final session on *Advanced Dialogue System Architecture*. All of them deal with the use of advanced statistical models to represent all or part of the dialogue framework, from natural language understanding (NLU) to dialogue manager (DM) to natural language generation (NLG).

The chapter by Stefan Ultes, Juliana Miehle and Wolfgang Minker, "On the Applicability of a User Satisfaction-Based Reward for Dialogue Policy Learning", concerns the search for a good dialogue policy that uses reinforcement learning. Rather than using objective criteria such as task success for reward modelling, they propose to use the subjective, but very realistic measure of user satisfaction. This measure is shown to be very successful when used on a user simulation.

The chapter by Michael Wessel, Girish Acharya, James Carpenter and Min Yin, "OntoVPA—An Ontology-Based Dialogue Management System for Virtual Personal Assistants", concentrates on the DM part of the dialogue architecture. The DM structure here is designed to address the issues of dialogue state tracking, anaphora and co-reference resolution by using an ontology. The ontology and its accompanying rules are used for both parts of the DM, state tracking and actions (response generation). They also show that the use of this architecture allows for easier creation of a dialogue system in a new domain, which is one of the most pressing concerns facing dialogue architecture research at present.

The chapter by Manex Settas, María Inés Torres and Arantza del Pozo, "Regularized Neural User Model for Goal-Oriented Spoken Dialogue Systems", presents a neural network approach for user modelling that exploits an encoder–decoder bidirectional architecture with a regularization layer for each dialogue act. The chapter deals with the data sparsity issue and presents results on the Dialogue State Tracking Challenge 2 (DSTC2) dataset.

The chapter by Robin Ruede, Markus Müller, Sebastian Stüker and Alex Waibel, "Yeah, Right, Uh-Huh: A Deep Learning Backchannel Predictor", looks at the acoustic cues of backchanneling. Departing from handwritten rules to represent the variety of backchannel cues, they use neural networks extended with long short-term memory (LSTM) networks to achieve better performance. Performance is further enhanced by the use of linguistic cues since these are often present along with the acoustic ones. Their use of this information in a word2vec set-up further increased accuracy.

Pittsburgh, USA
Orsay, France
March 2018

Maxine Eskenazi
Laurence Devillers
Joseph Mariani

Contents

Part I
Chatbots and Conversational Agents

Building Rapport with Extraverted and Introverted Agents

Jacqueline Brixey and David Novick

Abstract Psychology research reports that people tend to seek companionship with those who have a similar level of extraversion, and markers in dialogueshow the speaker's extraversion. Work in human-computer interaction seeks to understand creating and maintaining rapport between humans and ECAs. This study examines if humans will report greater rapport when interacting with an agent with an extraversion/introversion profile similar to their own. ECAs representing an extrovert and an introvert were created by manipulating three dialogue features. Using an informal, task-oriented setting, participants interacted with one of the agents in an immersive environment. Results suggest that subjects did not report the greatest rapport when interacting with the agent most similar to their level of extraversion.

Keywords Virtual agents · Dialogue systems · Rapport

1 Introduction

People often seek companionship with those who have a personality similar to their own [11]. There is evidence that personality types are borne out in dialogue choices [1]. Humans are uniquely physically capable of speech and tend to find spoken communication as the most efficient and comfortable way to interact, including with technology [2]. They respond to computer personalities in the same way as they would to human personalities [10]. Recent research has sought to understand the nature of creating and maintaining rapport—a sense of emotional connection—when communicating with embodied conversational agents (ECAs) [2, 8, 14].

Successful ECAs could serve in a number of useful applications, from education to care giving. As human relationships are fundamentally social and emotional, these

J. Brixey (✉)
USC Institute for Creative Technologies, Los Angeles, CA 90094, USA
e-mail: brixey@ict.usc.edu

D. Novick
University of Texas at El Paso, El Paso, TX 79968, USA
e-mail: novick@utep.edu

M. Eskenazi et al. (eds.), *Advanced Social Interaction with Agents*, Lecture Notes in Electrical Engineering 510, https://doi.org/10.1007/978-3-319-92108-2_1

qualities must be incorporated into ECAs if human-agent relationships are to feel natural to users. Research has been focused on the development and maintenance of rapport felt by humans when interacting with an ECA [11] and in developing ECA personalities [3]. However, questions remain as to which agent personality is the best match for developing rapport in human-ECA interactions.

In this study, two agents representing an extravert and an introvert were created by manipulating three dialogue features. Using a task-oriented but informal setting, participants interacted with an agent in an immersive environment, and then responded to a 16-question rapport survey.

This study specifically seeks to answer three questions: (1) Will all subjects establish a high-level of rapport with the extraverted agent? (2) Will extraverted subjects matched with an introverted agent show the lowest level of rapport due to mismatched personalities? (3) If all subjects establish rapport with the agent, will subjects report the highest level of rapport when interacting with the agent whose extraversion level matches that of the subject?

2 Extraversion and Introversion in Dialogue

Personality is a patterned set of organized traits and externalized behaviors that are permanent or slowly changing that influence a person's attitudes and emotional responses. Personality is considered a universal phenomenon of human psychology [16]. One prominent structure in the literature for personality traits and dimensions is the Five Factor Model (FFM). One factor in the FFM is extraversion-introversion, which is a consistent and salient personality dimension [16]. Extraversion is often displayed through "energetic" behavior and an outgoing and sociable attitude, while introversion is marked by quietness and seeking of solitude [5].

2.1 Extraversion in Dialogue

Personality markers in speech are dialogue acts that are associated with personality dispositions. Extraverted and introverted personalities are revealed based on expression of semantically equivalent but emotionally distinct phrases [2]. Extraverts tend to speak louder and talk more than listen [16]. Extraverts reflect their sociability by referring to other people, using more positive language, and saying more per conversation turn [15]. Introverts tend to use fewer positive words and give fewer compliments [10].

Social dialogue is talk in which interpersonal goals are the primary purpose, while task goals are secondary, and are used to build rapport and trust [1]. Social dialogue significantly increased trust for extraverts in ECA interactions but made no difference for introverts [2]. In contrast, introverts dislike social dialogue and prefer task-only dialogue [1].

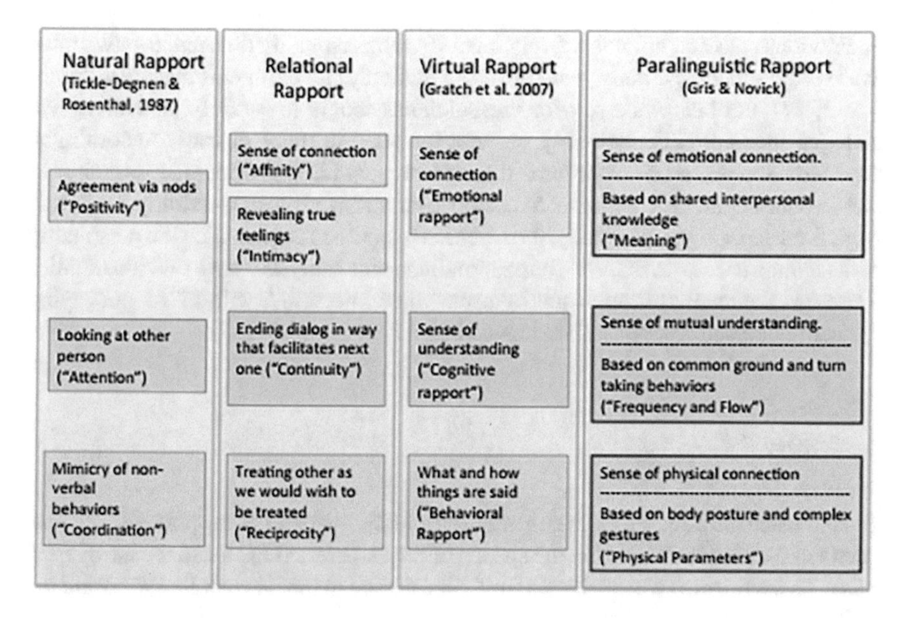

Fig. 1 Paralinguistic model of Rapport [14]

Extraverted speech correlates negatively with concreteness [6], achieved by adding subordinate clauses instead of starting a new sentence. This increases the number of noun and verb phrases, a sentence's length and syntactic complexity [15], and overall word count. Word count is an important surface feature for determining extravert dialogue [10]. Extraverts also tend to use a less robust vocabulary and produce less formal sentences [10]. For this study, formality is measured by [9]:

F = (noun frequency + adjective frequency + preposition frequency + article frequency − pronoun frequency − verb frequency − adverb frequency − interjection frequency + 100)/2

2.2 *Rapport with ECAs*

Rapport is the harmonious feeling experienced when forming an emotional connection with another person. It has profound effects on human relationships, from customer satisfaction and loyalty [7], to student success [18]. The incorporation of social and emotional qualities into the framework of the ECA facilitates the building and maintenance of human-ECA relationships, and rapport provides a basis for long-term human-ECA relationships. The paralinguistic rapport model used in this study measures rapport in three dimensions: a sense of emotional connect, a sense of mutual understanding, and a sense of physical connection [14] (Fig. 1).

Previous research did not actively quantify the personality difference between an extraverted and introverted agent, rather qualitatively created agents that are different [4, 13, 17]. Further, while previous research has sought to quantify personality via dialogue markers [11], formality was not included in those metrics. Accordingly, this study sought to select features that can be quantified and include formality in that measurement. Previous research has also emphasized agent likeability, while this work focuses on rapport. Finally, the research to date has not matched the personality of the human user to an ECA of a corresponding personality, nor have previous studies addressed the level of rapport felt by users when interacting with ECAs portraying a specific personality type via dialogue choices.

3 Methods

To determine the effect of ECA extraversion and introversion on rapport, we manipulated three dialogue features in the speech of the extraverted ECA and the introverted ECA. These features were selected for their salience in the relevant literature and the ability to measure each feature.

1. Positivity, as measured by the linguistic inquiry and word count (liwc.wpengine.com). This measurement was set to be higher in extraverted dialogue.
2. Word count, as measured as the total number of words per scene. A higher word count tends to be a marker of extraversion.
3. Formality, as measured by the formality formula [9]. This feature tends to be higher for introverts.

Participants in the study interacted with the agent in a game, "Survival on Jungle Island" [8], an interactive game where participants find themselves stranded on a remote island with the agent and work to try to survive and escape off the island. The game comprises ten scenes, which are interactions centered on a topic, such as family, or a task, for example trying to start a fire. Figure 2 shows the agent in a scene from the jungle game. The user interacted with the agent via spoken dialogue, and each scene was written to incrementally disclose information about the agent and to request similar information from the user. Since intimacy building dialogues do not change regardless of the previous decisions made by the user, all users have an equal opportunity to build rapport with the agent.

A script for the ECA was handwritten prior to manipulations [8] to ensure that the same content was given by the ECA to all participants. An example dialogue from Scene 2 is shown in Table 1. In this scene, the participant has just woken up on the beach and meets the agent.

All dialogues followed the same story arc and revealed the same information about the agent. While no metric was set that any of the factors had to be a specific ratio between the two agents, our objective was generally to make extraverted word count and positivity higher than that of introverted, and introverted formality higher

Fig. 2 The agent in a scene in the jungle game

Table 1 Example of extraverted ECA and introverted ECA dialogue

Line	Extraverted ECA	Introverted ECA
1	Oh, I'm so sorry, where are my manners, my name is Lina.	My name is Lina
2	How about you, what's your name?	What is your name?
3	Great to meet you, I'm so glad I'm not alone anymore!	It is very nice to meet you!
4	Well I've been on this island for a few days now.	I landed here seven days ago

than that of the extravert. Table 2 presents total word count, formality, and positivity for each scene. The extravert word count was, on average, 63% higher 5 than the introvert word count. The extravert formality was about 15% lower than the introvert formality. And the extravert positivity was about 100% higher than the introvert positivity.

In total, we recruited 59 subjects, 9 females and 50 males; four data points from the 59 recorded were discarded due to technical difficulties. After consenting, all participants completed a standard Myers-Briggs personality assessment to determine extraversion or introversion [12]. Participants were matched to interact with an agent at random. There were 15 participants in the introverted agent extraverted user (IE) group, 15 in the extraverted agent-introverted user (EI) group, 14 in the introverted agent-introverted user (II) group, and 11 in the extraverted agent-extraverted user (EE) group. The physical appearance and behavior of the ECA was the same for all scenarios; only the ECA's language was changed.

Table 2 A summary of word count, formality, and positivity for the respective agents in the Jungle Game

Scene	Extravert			Introvert		
	Word count	Formality	Positivity	Word count	Formality	Positivity
1	196	33	4.08	128	40	2.34
2	163	43	3.07	65	57	1.54
3	353	38	3.97	239	41	3.77
4	218	40	4.13	136	51	3.68
5	278	46	6.47	152	54	5.26
6	471	37	4.25	301	59	3.65
7	187	41	2.67	129	48	1.55
8	13	49	0	6	50	0
9	11	50	0	5	51	0
10	13	46	15.38	6	47	0
Total	1903	423	44.02	1167	498	21.79
Avg	190.30	42.30	4.40	116.70	49.80	2.17
SD	152.58	5.49	4.32	100.26	6.16	1.87

Experiments lasted for 30 min, with approximately 15 min of interaction with the agent. Following the interaction, participants filled out a 16-question rapport survey, with five questions related to emotional rapport, five related to cognitive rapport, and six questions dealing with behavior rapport. All survey questions were answered on a five-point Likert scale, with high agreement at 5. A manipulation check was also included on the rapport survey, where participants were asked, "How extraverted do you think the agent was?" and ranked the agent on a scale from 1 (very introverted) to 5 (very extraverted). Participants were also asked a free response question, "Do you prefer interacting with extraverts usually?" and to provide any comments about the experiment.

4 Results

The results for the research questions will be presented first, followed by a general discussion of the results. For the reported t-tests, the distributions of mean rapport scores for each experimental condition appear to be normal; Table 3 reports the Anderson-Darling p values, where $p > 0.05$ indicates that a normal distribution cannot be rejected. That is, all four conditions have acceptably normal distributions.

Table 3 Anderson-Darling test for normality of distributions of rapport scores

Condition	Anderson-Darling p Value
EE	0.62
II	0.44
IE	0.89
EI	0.18

4.1 Research Questions

(1) *Will all subjects establish a high level of rapport with the extraverted agent?*

The results indicate that all participants established a higher level of rapport with the extraverted ECA than with the introverted ECA, as can be seen in Fig. 3, which presents the averages and standard deviations of reported rapport based on answers to the 16-question survey. Overall, the EE group had the highest average level of rapport, followed by the IE group. Contrary to our hypothesis (and psychology literature), the introverted participants who interacted with the introverted agent (II) reported the lowest average levels of rapport overall and in each of the three subsections of rapport.

(2) *Will extraverted subjects matched with an introverted agent show the lowest level of rapport due to mismatched personalities?*

The results indicate that this is not true; instead, introverts reported the lowest level of rapport when interacting with the introverted agent (see Fig. 3). While extraverts did report a lower level of rapport with the introverted ECA than with 7 the extraverted ECA, the II group reporting the lowest level of rapport suggests that mismatched personalities do not predict the lowest level of rapport.

Table 4 presents the results from conducting t-tests on the data set. In general, the emotional and cognitive dimension did not show meaningful differences across the experimental conditions. However, there appear to be clear results for the behavioral dimension.

The result of $p < 0.05$ for behavioral rapport for the EE versus EI groups suggests that extraverts developed lower levels of rapport in the mismatched interaction. The lack of a significant result in the t-test for II versus IE (a comparison of the impact on rapport for mismatching for introverted participants) suggests that a mismatched ECA personality did not have a significant effect on the development of rapport for introverted subjects. These results indicate that introverted and extraverted participants do not have the same reaction in building rapport with the ECA as a function of ECA extraversion/introversion.

(3) *Will subjects report the highest level of rapport when interacting with the agent whose extraversion level matches that of the subject?*

The results suggest that extraverts did report the highest level of rapport with the extraverted ECA, but the II group showed the lowest average for rapport for all the groups (see Fig. 3), which was not consistent with the psychology research on

OVERALL

PARTICIPANT		AGENT: Introvert	AGENT: Extravert
	Introvert	**3.16** *1.37*	**3.35** *1.33*
	Extravert	**3.25** *1.43*	**3.62** *1.28*

BEHAVIORAL

PARTICIPANT		AGENT: Introvert	AGENT: Extravert
	Introvert	**3.13** *1.38*	**3.38** *1.33*
	Extravert	**3.21** *1.35*	**3.71** *1.29*

COGNITIVE

PARTICIPANT		AGENT: Introvert	AGENT: Extravert
	Introvert	**3.87** *1.23*	**3.89** *1.31*
	Extravert	**3.86** *1.28*	**4.04** *1.20*

EMOTIONAL

PARTICIPANT		AGENT: Introvert	AGENT: Extravert
	Introvert	**2.49** *1.12*	**2.79** *1.12*
	Extravert	**2.69** *1.39*	**3.11** *1.27*

Fig. 3 The averages (bold) and standard deviations (italicized) of the full data set

Table 4 The results of a t-test on the data set, ** indicates $p < 0.05$, * indicates $p < 0.10$

Condition	Overall	Emotional	Cognitive	Behavioral
EE/II	0.09*	0.14	0.57	0.02**
EE/IE	0.22	0.31	0.54	0.08*
EE/EI	0.18	0.34	0.53	0.05*
II/IE	0.64	0.51	0.96	0.59
II/EI	0.72	0.58	1	0.81
IE/EI	0.91	0.97	0.95	0.77

personality that reported that humans prefer to interact with those who have a personality most similar to their own. The significant ($p < 0.05$) t-test result for behavioral rapport of the EE versus II groups indicates that introverts feel substantially less rapport on average when interacting with an agent of the same personality type than do extraverts.

Table 5 The effect sizes in the full set of data, * indicates $\theta > 0.60$, and ** indicates $\theta > 0.80$

Effect size	Overall	Emotional	Cognitive	Behavioral
EE/II	0.84**	0.73*	0.24	1.19**
EE/IE	0.47	0.41	0.21	0.71*
EE/EI	0.65*	0.46	0.29	0.98**
II/IE	0.31	0.34	0.04	0.38
II/EI	0.13	0.19	0.01	0.12
IE/EI	0.15	0.11	0.05	0.25

4.2 *Effect Size*

We examined effect size to provide insight into the size of the difference between the various groups' mean rapport, shown in Fig. 3. We calculated effect size θ using the standardized mean difference between populations to provide additional perspective on the differences between the various groups' mean reported rapport scores (see Table 5). A larger absolute value for effect size indicates a stronger effect and complements the findings of the t-test results from Table 4 by giving insight into the strength of the statistical claim. We found an exceptionally strong effect of 0.84 for overall rapport when comparing the EE and II groups (effect > 0.8), disconfirming the expectation from psychology for our subjects in human-ECA interactions. The strong effect of 0.65 found in the EE versus EI comparison (effect > 0.8) suggests that extraverts will report lower levels of rapport when matched with an introverted agent. The lack of effect from the introverted participants in either agent-matching condition suggests that introverted participants do not feel more rapport with either agent and also indicates that extroverts in general report higher levels of rapport than introverted participants when interacting with an agent.

The results of the analysis of effect are consistent with the question posed to participants on the rapport survey, "Do you prefer interacting with extraverts usually?" Introverted participants reported mixed preference, with 40% preferring other introverts, 40% preferring extraverts, and the remaining subjects having no preference. However, 80% of extravert participants preferred other extraverts, with 8% preferring introverts and 12% with no preference. This result provides a new perspective on the psychology of introverts as well as on the dynamics of rapport in relation to extraversion, as it appears that introverts do not have the same expectations for their conversation partners, either human or ECA, as extraverts. Thus, the creation and maintenance of rapport with introverted participants will not be identical to that of extraverted participants.

Finally, all participants answered a manipulation check on the rapport survey, "How extraverted do you think the agent was?" Both the extravert subjects and the introvert subjects rated the extraverted agent as more extraverted than the introverted agent. However, these differences were small (see Table 6) and not statistically significant; overall, both extravert and introvert participants rated the extraverted agent

Table 6 Perceived extraversion, by subject and agent condition

Condition	Mean perceived extraversion	Difference
EE	4.45	0.38
EI	4.07	
IE	4.47	0.07
II	4.40	

as highly extraverted. Extraverted participants rated the introverted agent as being more introverted than did the introverted participants. Some responses from participants as to why they did not perceive the introversion of the agent were that the ECA initiated too many conversations and conversation topics, particularly those about herself. These responses match with descriptions that introverts prefer task dialogue to social dialogue [1] and to let others make decisions [10], features that were not incorporated into the present ECA interaction model. Another potential explanation is that the phrasing of the question might have led participants to expect the agent was intended to be extraverted.

5 Conclusion

Our study's results suggest that while there are instances in which the behavior of humans in human-ECA interactions is the same as those in human-human interaction, human-ECA interactions with introverted agents may not be one of these instances. Should only one agent personality be available to implement, our results suggest that the best option would be an extraverted personality, as both introverts and extraverts reported higher levels of rapport with an extroverted agent than introverted.

The study also found that introverts did not show preference for either agent. Future work should explore how to better relate to and interact with introvert participants. Future work could also determine if introvert participants develop more rapport with an agent during a longer or multi-session interaction.

This study suggests that personality type expressed dialogue can be felt by users, and this differentiation affects the behavior dimension of rapport, giving new 10 insights into the inner workings and potential manipulations of to increase rapport with ECAs. Finally, as only three dialogue features were manipulated to differentiate the two agents, future work will incorporate more and different variations of extravert versus introvert dialogue features.

References

1. Bickmore T, Cassell J (2005) Social dialogue with embodied conversational agents. In: Advances in natural multimodal dialogue systems. Springer, Netherlands, pp 23–54
2. Cassell J (2000) Embodied conversational agents. MIT press
3. Cassell J, Bickmore T (2003) Negotiated collusion: modeling social language and its relationship effects in intelligent agents. User Model User-Adap Inter 13(1–2):89–132
4. Chen Y, Naveed A, Porzel R (2010) Behavior and preference in minimal personality: a study on embodied conversational agents. In: International conference on multimodal interfaces and the workshop on machine learning for multimodal interaction. ACM
5. Eysenck SBG, Eysenck HJ, Barrett P (1985) A revised version of the psychoticism scale. Personality Individ Differ 6(1):21–29
6. Gill A, Oberlander J (2002) Taking care of the linguistic features of extraversion. In: Proceedings of the 24th annual conference of the cognitive science society
7. Gremler DD, Gwinner KP (2000) Customer-employee rapport in service relationships. J Serv Res 3(1):82–104
8. Gris I (2015) Two-way virtual rapport in embodied conversational agents. Doctoral dissertation, University of Texas at El Paso
9. Heylighen F, Dewaele J-M (2002) Variation in the contextuality of language: an empirical measure. Found Sci 7(3):293–340
10. Mairesse F et al (2007) Using linguistic cues for the automatic recognition of personality in conversation and text. J Artif Intell Res 30:457–500
11. Mairesse F et al (2007) PERSONAGE: personality generation for dialogue. Annu Meet Assoc Comput Linguist 45:496–503
12. Myers I, Myers P (2010) Understanding personality type. Nicholas Brealey Publishing, Gifts differing
13. Nass C et al (1995) Can computer personalities be human personalities? In: Conference companion on Human factors in computing systems. ACM
14. Novick D, Iván G (2014) Building rapport between human and ECA: A pilot study. In: International conference on human-computer interaction. Springer International Publishing
15. Scherer KR (1979) Personality markers in speech. Cambridge University Press
16. Scherer KR (1981) Speech and emotional states. Speech Eval Psychiatry:189–220
17. Van den Bosch K et al (2012) Characters with personality!. In: International conference on intelligent virtual agents. Springer, Berlin Heidelberg
18. Wilson JH, Ryan RG, Pugh JL (2010) Professor-student rapport scale predicts student outcomes. Teach Psychol 37(4):246–251

Automatic Evaluation of Chat-Oriented Dialogue Systems Using Large-Scale Multi-references

Hiroaki Sugiyama, Toyomi Meguro and Ryuichiro Higashinaka

Abstract The automatic evaluation of chat-oriented dialogue systems remains an open problem. Most studies have evaluated them by hand, but this approach requires huge cost. We propose a regression-based automatic evaluation method that evaluates the utterances generated by chat-oriented dialogue systems based on the similarities to many reference sentences and their annotated evaluation values. Our proposed method estimates the scores of utterances with high correlations to the human annotated scores; the sentence-wise correlation coefficients reached 0.514, and the system-wise correlation were 0.772.

Keywords Conversational systems · Automatic evaluation
Reference-based estimation

1 Introduction

The enormous cost of evaluating chat-oriented dialogue systems is one major obstacle to improve them. Previous work has evaluated dialogue systems by hand [1, 2], which is a common practice in dialogue research. However, such an approach not only requires a huge cost but it is also not replicable; i.e., it is difficult to compare a proposed system's scores with the previously reported scores of other systems.

As a first trial of substituting human annotations, Ritter et al. introduced BLEU, which is a reference-based automatic evaluation method widely used in the

H. Sugiyama (✉) · T. Meguro · R. Higashinaka
NTT Communication Science Laboratories, 2-4, Hikari-dai, Seika-cho, Souraku-gun, Kyoto, Japan
e-mail: sugiyama.hiroaki@lab.ntt.co.jp
URL: http://www.kecl.ntt.co.jp/icl/ce/member/sugiyama/

T. Meguro
e-mail: meguro.toyomi@lab.ntt.co.jp

R. Higashinaka
NTT Media Intelligence Laboratories, 1-1, Hikari-no-oka, Yokosuka-shi, Kanagawa, Japan
e-mail: higashinaka.ryuichiro@lab.ntt.co.jp

M. Eskenazi et al. (eds.), *Advanced Social Interaction with Agents*, Lecture Notes in Electrical Engineering 510, https://doi.org/10.1007/978-3-319-92108-2_2

assessment of machine-translation systems [3, 4]. They evaluate their dialogue systems on the basis of the appropriateness of each one-turn response for input sentences instead of whole dialogues. While such a reference-based evaluation methodology shows high correlations with human annotators in machine-translation, they reported that the reference-based approach fails to show high correlation with human annotations in the evaluation of chat-oriented dialogues. In machine-translation, since systems are required to generate sentences that have exactly the same meaning as the original input sentences, the appropriate range of the system outputs is so narrow that only one or just a few reference sentences are enough to cover them. On the other hand, in chat-oriented dialogues, since the appropriate range is likely to be much larger than in machine-translation, such a small number of references is likely to be insufficient.

Galley et al. proposed Discriminative BLEU (Δ BLEU), which leverages 15 references with manually annotated evaluation scores to estimate the evaluation of chat-oriented dialogue system responses [5]. Their method leverages *negative* references in addition to customary *positive* references in the calculation of BLEU and evaluates sentences that resemble negative references as inappropriate. This increases the correlation with human judgment up to 0.48 of Pearson's r when the correlation is calculated with 100 sentences as a unit; however, they also reported that sentence-wise correlation remained low:$r \leq 0.1$. The reason is probably that their method evaluates a sentence that is far from all the references as neutral rather than inappropriate.

We propose a *regression*-based approach that automatically evaluates chat-oriented dialogue systems by leveraging the distances between system utterances and a large number of positive and negative references. We expect our regression-based approach to appropriately evaluate sentences that are not similar to all of the references. We also gathered a larger scale of references than Galley's work and examined the effectiveness of the number of references over the estimation performance.

2 Multi-reference-Based Evaluation

This section explains how we gather positive/negative sentences by humans, consistently evaluate them among the annotators, and automatically estimate their evaluation scores.

2.1 *Development of Reference Corpus*

We developed a multi-reference corpus that contains both positive and negative reference sentences (responses to input sentences). To collect reasonable input-response pairs for automatic evaluation, first we collected utterance-like sentences as input sentences from the web and real dialogues between humans. To remove sentences that require understanding of the original contexts, human annotators rated the *comprehensibility scores* of the collected sentences (degrees of how the annotators can easily understand the situations of the sentences), and we randomly chose sentences with high comprehensibility scores.

After collecting the input sentences, 10 non-expert reference writers created response sentences that would satisfy users. To intentionally gather inappropriate responses, we designed the following two constraints of their creation: *character-length limitation* and *masked input sentences*. The character-length limitation, which narrows the available expressions, decreases the naturalness of the references. The limitation on masked input sentences means that the reference writers create responses using sentences whose words are partly deleted. For example, when we mask 60% of the words in the following sentence *What is your favorite subject?*, *What is *** *** ***?* or **** is *** favorite ***?* is shown to the reference writers. This enables us to gather response sentences that have irrelevant content to the original input sentences and simultaneously maintain the syntactic naturalness of the responses. In addition, to add other types of inappropriate sentences to the negative references, we gathered sentences that were generated by existing dialogue systems described in Sect. 3.1.

2.2 Evaluation of References

Human annotators evaluate reference sentences in terms of their naturalness as responses. In this work, we adopted the **pairwise winning rate** over all other references as an evaluation score of a reference sentence. If a sentence is judged to be more natural than all the other references, its evaluation score is 1; a sentence that is judged the least natural obtains an evaluation score of 0. Our preliminary experiment showed that if the evaluation scores are rated on a 7-point Likert scale, they tend to be either maximum or minimum; 45% were rated as 7 and 25% as 1. Hence it is difficult to determine the differences among the references by their scores. On the contrary, the winning rates vary broadly, and we can precisely distinguish differences among the references.

The drawback of the winning rate is its evaluation cost; the number of pairwise evaluations of N references is $N(N-1)/2$. However, pairwise evaluations for partly sampled pairs are reported to be satisfactorily accurate to maintain the winning rates [6].

2.3 Score Estimation Methods

Our method estimates a score of a pair of an input sentence and its response T. We considered the following three approaches in order to automatically evaluate system responses using the gathered pairs of input-references with human-annotated evaluation scores (pairwise winning rates).

Average of metrics (AM) This method outputs an estimation score of target sentence T with the average of sentence-wise similarities $s_{(T,R_m)}$ with top-M similar reference sentences R_m as follows:

$$E_{AM}(T) = \frac{\Sigma_{m=1}^{M} s_{(T,R_m)}}{M}. \quad (1)$$

This utilizes only the similarities with the references and resembles the approach for machine-translation. Since this assumes that only positive references are input, we just use manually created references without a masking constraint.

Weighted scores (WS) This method first calculates the sentence-wise similarities $s_{(T,R_m)}$ with top-M similar reference sentences R_m. This method also calculates the evaluated scores e_m (winning rates) of the top-M similar references. Then it outputs the average of the scores e_m of the top-N similar references weighted by the similarities $s_{(T,R_m)}$ as

$$E_{WS}(T) = \frac{\Sigma_{m=1}^{M} \{e_m \cdot s_{(T,R_m)}\}}{M}. \quad (2)$$

Regression This estimates the evaluation scores with regression models like Support Vector Regression (SVR) [7]. We used similarity metrics $s(T, R_n)$ for N references as features and trained the model with data $\mathcal{D} = (\mathbf{x}_i, e_i)_{i=1}^{N}$, where $\mathbf{x}_i = \{s(R_i, R_j)\}, j \in \{1, \ldots, N\}$. Here, $s(R_i, R_j)$ means a similarity score between reference R_i and R_j. We developed a regression model for each input sentence.

3 Experiments

First we gathered pairs of input and reference sentences with evaluation scores. Then, based on the references, we developed evaluation score estimators and examined the effectiveness of our multi-reference approach.

3.1 Settings

Input sentences We sampled input sentence candidates from a chat-oriented dialogue corpus as well as a Twitter corpus. Both corpora contain only Japanese sentences. The chat-oriented dialogue corpus consisted of 3680 one-to-one text-chat dialogues between Japanese speakers without specified topics [1]. From this corpus, we extracted input sentence candidates whose dialogue-acts were related to

Table 1 Statistics of gathered references for an input sentence. Human denotes number of manually created references and the others are automatically created references

Method	$N_c < 50$	$10 \leq N_c < 50$	$N_c < 10$	Sum
Human (no mask)	18	18	6	42
Human (30% mask)	6	6	2	14
Human (60% mask)	6	6	2	14
IR-status	10	0	0	10
IR-response	10	0	0	10
Rule	10	0	0	10
Sum	60	30	10	100

self-disclosure. From Twitter, we sampled sentence candidates that contain topic words, which were extracted from the top-10 ranked terms of Google trends in 2012 in Japan.[1]

To remove candidates that require the contexts of the original dialogues to be understood by the writers, we recruited two annotators who rated the *comprehensibility scores* of the sentence candidates on a 5-point Likert scale and only used sentences that received scores of 5 from both annotators as input sentences. For the following experiment, we used ten input sentences: five randomly sampled from the conversational corpus and five from the Twitter corpus. The number of input sentences may be small due to the cost of labeling as we describe in the next section.

Reference sentences and evaluations Using the ten selected input sentences, ten reference sentence writers (not the annotators for comprehensibility scores) created references. Each writer created seven reference sentences for each input sentence under two constraints: character-length limitation and masked input sentences. Table 1 shows the statistics of the gathered references. As a limitation of character length N_c, one reference writer created three sentences under the $N_c < 50$ condition (nearly free condition) that only limits excessively long references, three sentences under $10 \leq N_c < 50$ that forces writers to avoid overly simple references, and one sentence under $N_c < 10$ that forces writers to produce such simple references as *I guess so* or *That sounds good.*

The following are the details of the masked input sentences. For all input sentences, six writers created references for them without masks, two writers created references for 30% masked input sentences, and two writers created references for 60% masked sentences. We randomly assigned the input sentences to writers who imagined the masked terms and created references. They wrote 70 references for each input sentence: 42 sentences without masks, 14 with 30% masked, and 14 with 60% masked (Table 1).

In addition to the manually created references, we gathered 30 possibly negative reference sentences that were generated by the following two retrieval-based generation methods, *IR-status* and *IR-response* [3], and one rule-based generation

[1]https://www.google.co.jp/trends/topcharts#date=2012.

Table 2 Examples of input sentences, reference sentences and their winning rates

Input sentence	References	Sources	Winning rates
I don't like Disneyland when it's very crowded... そして、ディズニーランドの大混雑も苦手です…。	It's so insane when everyone starts dashing at the same time as the gates open. 開門と同時に皆が走り出す光景って、何度見てもぞっとしますよね。	Human 0%	0.96
	Oh, it was really bad.. あらら、それは大変でしたね。	Human 30%	0.43
	I'm surprised that anyone would have such a pet. 飼っている人がいると聞いてびっくりです。	Human 60%	0.01
	Yeah, I agree! あーあたしも！	IR-status	0.81
	It's so crowded! 大混雑だな！	IR-response	0.29
	Yes, I go to Disneyland over ten times a year. はい、ディズニーには年に 10 回は行きます。	Rule	0.20
I just checked my iTunes, and I know all of my songs are from animated movies, games, vocaloids, voice actors and audio dramas of comics. iTunes に入ってるの確認したらアニソンとゲーソンとボカロと声優さんとドラマ CD だらけだった	You must really like anime and games! アニメとかゲームが好きなんだね！	Human 0%	0.95
	Don't you listen to rocks or western music? ロックや洋楽は聞かないんですか？	Human 30%	0.88
	Whet is your favorite year of it? 一番あたりだった年はいつですか？	Human 60%	0.15
	That's normal for myself. いつものわたしである	IR-status	0.26
	Vocaloids and anime songs lol. ボカロとアニソンだね w	IR-response	0.43
	What anime songs do you like? アニソンは何が好きなんですか？	Rule	0.43

method, *Rule* [1]. *IR-status* retrieves reply posts whose associated source posts most closely resemble the input user utterances. The *IR-response* approach is similar to the *IR-status*, but it retrieves the reply posts that most closely resemble the input user utterances. *Rule* represents a rule-based conversational system that uses 149,300 rules (pattern-response pairs) written in AIML [8] and retrieves responses whose associated patterns have the highest word-based cosine similarity to the input sentence. Each method generated ten reference sentences for each input sentence.

After the reference collection, two human evaluators annotated the winner of each reference pair in terms of its *naturalness as a response*. With 100 references for each input sentence, they annotated 4,950 pairs for each input sentence. Table 2 shows examples of the input sentences and the references with their winning rates.

Estimation procedure We compared the three methods described in Sect. 2.3 with smoothed BLEU that calculates BLEU over multi-references (m-BLEU)[2] and ΔBLEU [5]. All the estimations were conducted through the leave-one-out method; i.e., the methods estimated the evaluation scores for each reference sentence using the other 99 references. The parameters of the methods are experimentally determined. We used 3 for the M of AM (Average of metrics) and WS (Weighted scores), SVR with RBF-kernel, and C=5. Similarity metrics s used in AM, WS, and Regression are either sentence-BLEU (BLEU), RIBES [9], or Word Error Rate (WER). Here,

[2] We used NIST geometric sequence smoothing, which is implemented in nltk (Method 3).

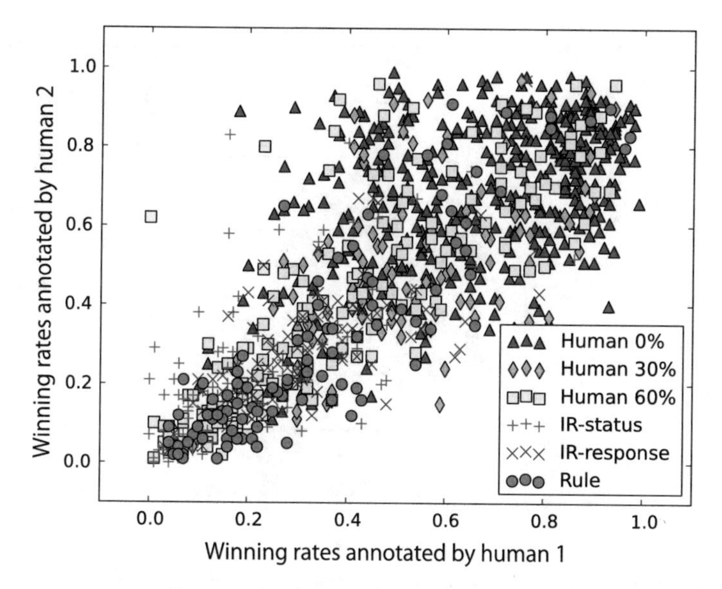

Fig. 1 Distribution of annotated winning rates between annotators

WER, which is calculated as normalized Levenshtein distance NL to a reference sentence, is converted to a similarity with either $WER{=}1-NL$ (ranges from 0 to 1) or $WER{=}1-2NL$ (-1 to 1).

3.2 *Analysis of Annotated Evaluations*

Before the experiments, we performed a brief analysis of the manually annotated evaluation scores (winning rates). Figure 1 shows their distribution between the annotators. They are broadly distributed along the whole range of 0-1. The manually created references (red triangles, orange diamonds, and yellow squares) were evaluated as more natural than the system-generated references. Comparing the system-generated references, those generated from the retrieval-based methods (*IR-status*: blue crosses, *IR-responses*: purple x marks) are gathered in the low or middle winning rates, and those generated from *Rule* (green circles) are distributed along the whole range. This shows that *Rule* generated references with the same appropriateness as the manually created ones when the rules correctly matched the input sentences. Pearson's correlation coefficient between the human evaluators was 0.783. Figure 1 also shows that the references with low winning rates show stronger correlations since the points of lower left corner gather at y = x. This result indicates that the negative input-response pairs are consistent between the evaluators, but the positive pairs are somewhat different probably because negative ones can be checked with violation of some criteria such as Grice's maxims [10].

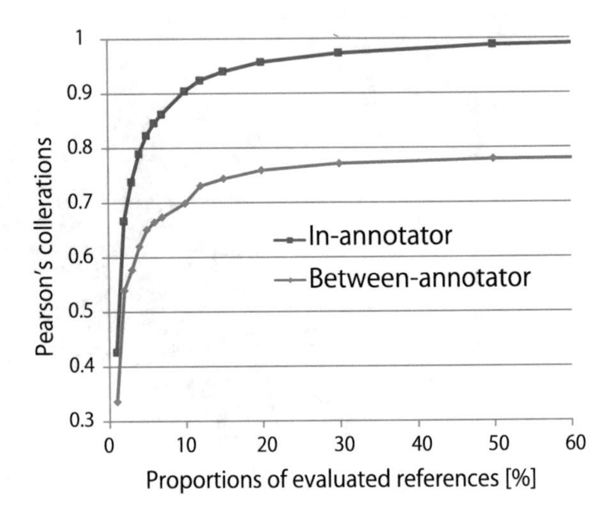

Fig. 2 Annotated pairwise propotions versus correlations

Figure 2 shows the variation of the pearson's correlation in- and between-evaluators over the rate of the evaluated pairs. We obtain the winning rates from partially sampled pairwise evaluations. The increase of the coefficients become slow around 12% of the evaluation rates. With our 100 references, 600 pairwise evaluations are enough to obtain the winning rates with high correlation coefficients ($r = 0.924$ in in-annotator condition and $r = 0.731$ in between-annotator condition) with the true rates.

3.3 *Results*

Figure 3 shows the correlation coefficients of the combination of the sentence similarity metrics and the proposed methods. SVR with WER (using the range from -1 to 1) shows the highest correlation ($r = 0.514$). Among the AM (Average of metrics) methods that leverage only the positive references, WER (-1 to 1) shows the highest correlation but still has lower correlations ($r = 0.399$) than those that leveraged the negative references. We calculated these scores using human 2 annotations, but it does not differ from the scores using human 1 annotations.

Figure 4 shows the relations between the number of references and the correlations of SVR with WER and AM with WER. With fewer references, AM shows higher correlations than SVR, because it requires training samples for accurate estimations, while AM can output reasonable estimations even with just one reference. The SVR performance becomes higher than AM with over 25 references and continues to improve. This indicates that both of our regression-based approach and the large size of references are keys to estimate the scores with high correlation.

Figure 5 shows the system-wise evaluation scores between manual annotation and SVR estimation. Each point is calculated as the means of ten scores; each score is

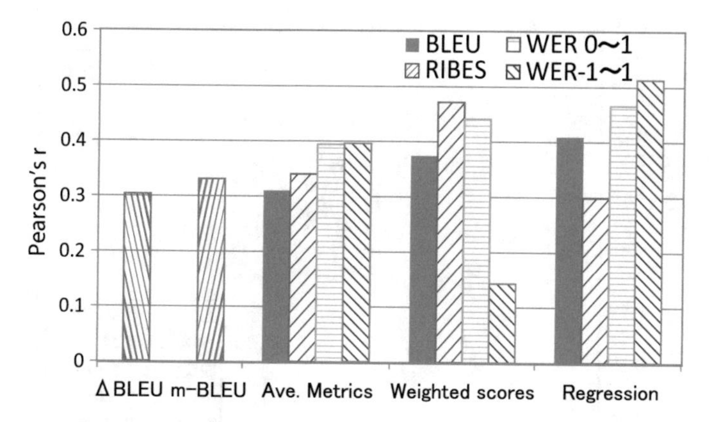

Fig. 3 Correlation between annotated (human 2) and estimated scores

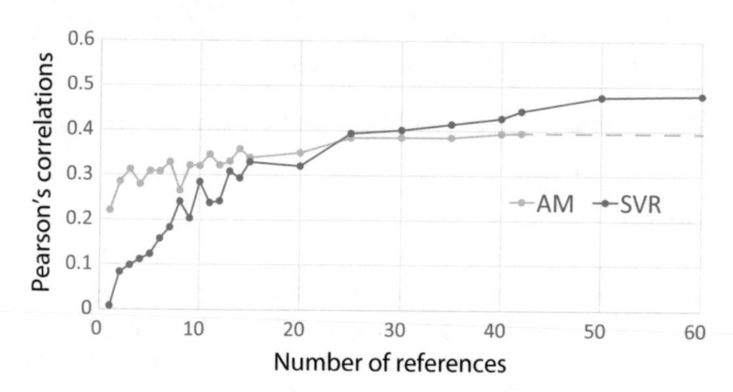

Fig. 4 Correlations over number of references

sampled from the estimated scores of an input-reference pair whose references are associated with certain generation methods (e.g., human 30% mask or *Rule*). The scores are highly correlated with Pearson's $r = 0.772$. Figure 5 illustrates that the references generated from human 60% mask, *Rule*, and *IR-status* are estimated with higher scores than the manual scores. This is because most of the low-evaluated references of human 60% mask and *Rule* have correct grammar but wrong contents and are barely distinguished with WER that only considers edit counts. *IR-status* has many expressions that did not appear in other references, such as *lol* (*www* in Japanese) and emoticons like *:-)*. The differences between these expressions and the references are difficult to evaluate with WER and BLEU because they depend on word matching. This problem may be solved using character N-grams and the proportions of the character types as regression features.

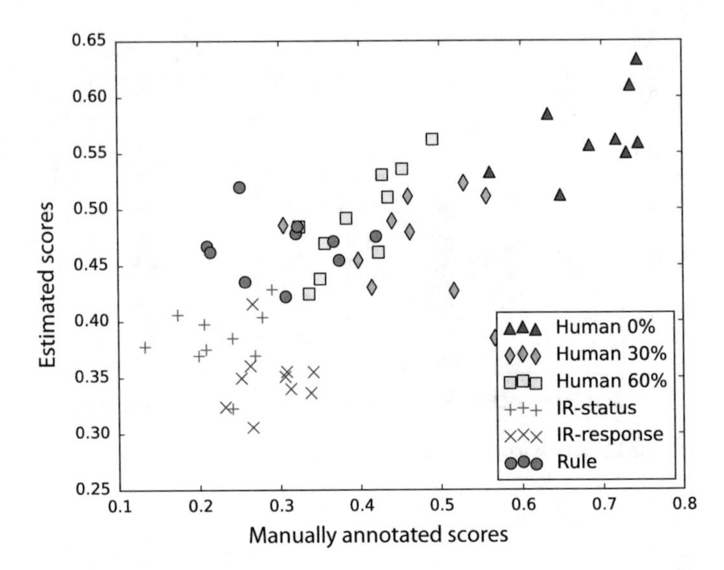

Fig. 5 System-wise comparison of annotated and estimated evaluation scores

4 Conclusion

We proposed a regression-based evaluation method for chat-oriented dialogue systems that appropriately leverages many positive and negative references. The sentence-wise correlation coefficient between our proposed and human annotated scores reached 0.514 and the system-wise correlations were 0.772. These scores are significantly higher than the previous methods such as delta-BLEU that define evaluation scores with sentences similarities between system output and reference sentences. Our results indicate that both of our regression-based approach and the large size of references are keys to estimate the scores with the high correlation. The limitation of our work are the small number of inputs and the huge cost of winning rates. We are planning to large-scale input-reference pairs with Likert scale evaluations to examine the effectiveness of our approach and the differences between winning rates and Likert scales.

References

1. Higashinaka R, Imamura K, Meguro T, Miyazaki C, Kobayashi N, Sugiyama H, Hirano T, Makino T, Matsuo Y (2014) Towards an open-domain conversational system fully based on natural language processing. In: Proceedings of the 25th international conference on computational linguistics, pp. 928–939

2. Sugiyama H, Meguro T, Higashinaka R, Minami Y (2014) Open-domain utterance generation using phrase pairs based on dependency relations. In: Proceedings of spoken language technology workshop, pp. 60–65
3. Ritter A, Cherry C, Dolan W (2011) Data-driven response generation in social media. In: Proceedings of the 2011 conference on empirical methods in natural language processing, pp. 583–593
4. Papineni K, Roukos S, Ward T, Zhu W (2002) BLEU: a method for automatic evaluation of machine translation. In: Proceedings of the 40th annual meeting on association for computational linguistics, pp. 311–318
5. Galley M, Brockett C, Sordoni A, Ji Y, Auli M, Quirk C, Mitchell M, Gao J, Dolan B (2015) Delta-BLEU: a discriminative metric for generation tasks with intrinsically diverse targets, pp. 445–450
6. Sculley D (2009) Large scale learning to rank. In: Proceedings of NIPS 2009 workshop on advances in ranking, pp. 1–6
7. Smola AJ, Sch B, Schölkopf B (2004) A tutorial on support vector regression. Statist Comput 14(3):199–222
8. Wallace RS (2004) The anatomy of A.L.I.C.E. ALICE Artificial Intelligence Foundation, Inc.
9. Isozaki H, Hirao T, Duh K, Sudoh K, Tsukada H (2010) Automatic evaluation of translation quality for distant language Pairs. In: Proceedings of the conference on empirical methods on natural language processing, pp. 944–952
10. Grice HP (1975) Logic and conversation. In: Syntax and semantics. 3: speech acts, pp. 41–58

What Information Should a Dialogue System Understand?: Collection and Analysis of Perceived Information in Chat-Oriented Dialogue

Koh Mitsuda, Ryuichiro Higashinaka and Yoshihiro Matsuo

Abstract It is important for chat-oriented dialogue systems to be able to understand the various information from user utterances. However, no study has yet clarified the types of information that should be understood by such systems. With this purpose in mind, we collected and clustered information that humans perceive from each utterance (perceived information) in chat-oriented dialogue. We then clarified, i.e., categorized, the types of perceived information. The types were evaluated on the basis of inter-annotator agreement, which showed substantial agreement and demonstrated the validity of our categorization. To the best of our knowledge, this study is the first to clarify the types of information that a chat-oriented dialogue system should understand.

Keywords Dialogue system · Chat-oriented dialogue · Perceived information

1 Introduction

Dialogue systems can use preselected knowledge about a specific task for understanding user utterances in task-oriented dialogue [2, 11], but this framework cannot be used in chat-oriented dialogue because it does not have (or at least seems not to have) a clear information structure.

Current chat-oriented dialogue systems interpret a user utterance by converting it into understanding results such as: keywords (approximating the focus or important information of a dialogue) [4], dialogue acts (for determining user intentions) [8], predicate argument structures (for understanding events) [4], emotions (for carrying

K. Mitsuda (✉) · R. Higashinaka · Y. Matsuo
NTT Media Intelligence Laboratories, Nippon Telegraph and Telephone Corporation,
1-1 Hikarinooka, Yokosuka, Japan
e-mail: mitsuda.ko@lab.ntt.co.jp

R. Higashinaka
e-mail: higashinaka.ryuichiro@lab.ntt.co.jp

Y. Matsuo
e-mail: matsuo.yoshihiro@lab.ntt.co.jp

M. Eskenazi et al. (eds.), *Advanced Social Interaction with Agents*, Lecture Notes in Electrical Engineering 510, https://doi.org/10.1007/978-3-319-92108-2_3

out an action fitting a user's emotion) [7], and user attributes (for personalizing a response) [6]. Although such information is used as understanding results of dialogue systems, it is not clear whether these types of information are sufficient for the understanding of chat-oriented dialogue systems.

To investigate what sorts of information chat-oriented dialogue systems must understand, we focused on the information that humans perceive from each utterance in a dialogue. We call such information "perceived information." Tasks that relate to perceived information have been emerging recently; such as conversation entailment [13, 14], irony detection [5, 9], and document enrichment [15, 16]. However, they only focus on a specific type of perceived information and do not provide an overall picture or categorization of perceived information.

In this paper, we report on the data collection and analysis of perceived information. We collected many instances of perceived information written by multiple annotators. We then had other annotators manually cluster the same type of instances for analyzing what types of perceived information exist. To evaluate the clustering results, we tested the inter-annotator agreement in annotating the types of perceived information. Through this analysis, we clarified the kinds of information that dialogue systems should understand from utterances in chat-oriented dialogue.

2 Collection of Perceived Information

Here, we describe how we collected the perceived information in chat-oriented dialogue. First, we prepared dialogues for which the perceived information would be written down. Second, we collected perceived information by having multiple annotators write the perceived information for each utterance in the dialogues.

2.1 Preparation of Dialogues

The choice of dialogues is important because they will be the source of the perceived information. For collecting general and various perceived information, we used a Japanese chat-oriented dialogue corpus collected by Higashinaka et al. [4], containing 3,680 dialogues between two people on various topics. We randomly selected 30 dialogues, which totaled 1,103 utterances.

2.2 Collection Procedure

We collected perceived information (hereafter, we refer to it as *PerceivedInfo*). We defined *PerceivedInfo* as the information that humans can generally understand from an utterance in dialogue even though the information may not be explicit.

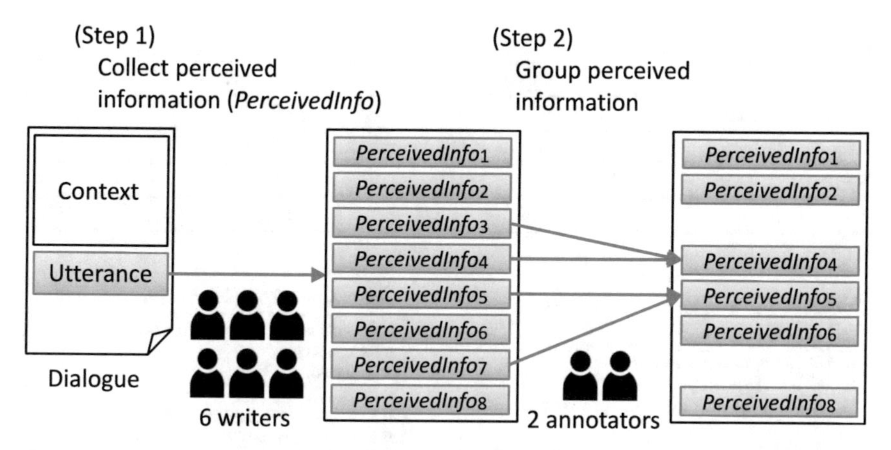

Fig. 1 Data collection and grouping of perceived information

The procedure of data collection is two-fold, as illustrated in Fig. 1. It is similar to the data-collection procedure of conversation entailment, where entailed information is collected from conversational data [13] and contradictory event pairs from propositional data [10].

(Step 1) First, annotators wrote *PerceivedInfo* as a natural sentence with regard to each utterance as a target in the dialogue. They wrote *PerceivedInfo* for all utterances in the dialogue in order from first to last. They could only use the context before the target utterance for writing *PerceivedInfo*; and the context after the target could not be used. They were instructed to write one or more *PerceivedInfo* for each utterance. We also told them that *PerceivedInfo* could not be a simple paraphrase of an utterance, complementation of omitted words, or information that is trivial on the basis of common sense, e.g., "We eat bread" or "Bread is made from flour." We made it mandatory that each *PerceivedInfo* had a predicate and an argument so that the proposition would be meaningful. We also made it mandatory that a *PerceivedInfo* should only be a single piece of information; thus, multiple pieces of information were divided into multiple *PerceivedInfo*. Six annotators worked independently in this step.

(Step 2) Second, duplicated or semantically similar *PerceivedInfo* for each utterance were grouped for the later process of clustering. Annotators who were not writers in Step 1 initially grouped *PerceivedInfo* independently and consulted one another to come up with the final results. If the content of multiple *PerceivedInfo*, such as content word, modality, tense, and negation were the same, they were grouped as the same *PerceivedInfo*. A representative *PerceivedInfo* written with the simplest wording was selected from each group. The representative *PerceivedInfo* was used in the next process of clustering.

Table 1 Amount of collected perceived information

	Dialogue	*PerceivedInfo*
Unique sentence	1,094	8,794
Unique word	1,596	3,740
Sentence	1,257	11,533
Word	10,856	116,413

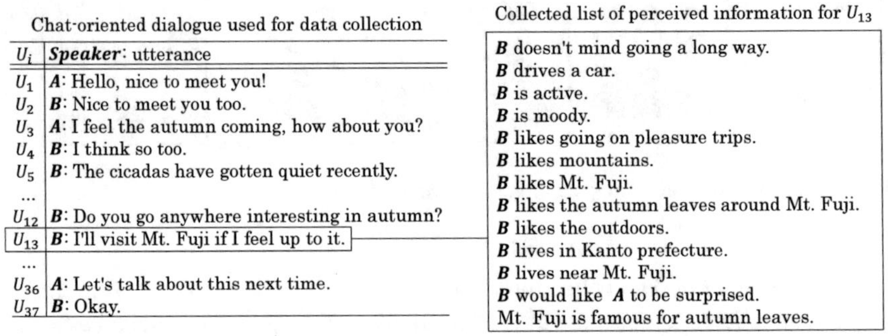

Chat-oriented dialogue used for data collection

U_i	*Speaker*: utterance
U_1	***A***: Hello, nice to meet you!
U_2	***B***: Nice to meet you too.
U_3	***A***: I feel the autumn coming, how about you?
U_4	***B***: I think so too.
U_5	***B***: The cicadas have gotten quiet recently.
...	
U_{12}	***B***: Do you go anywhere interesting in autumn?
U_{13}	***B***: I'll visit Mt. Fuji if I feel up to it.
...	
U_{36}	***A***: Let's talk about this next time.
U_{37}	***B***: Okay.

Collected list of perceived information for U_{13}

B doesn't mind going a long way.
B drives a car.
B is active.
B is moody.
B likes going on pleasure trips.
B likes mountains.
B likes Mt. Fuji.
B likes the autumn leaves around Mt. Fuji.
B likes the outdoors.
B lives in Kanto prefecture.
B lives near Mt. Fuji.
B would like ***A*** to be surprised.
Mt. Fuji is famous for autumn leaves.

Fig. 2 Example of chat-oriented dialogue and perceived information for utterance

We recruited 12 annotators for Step 1 and two for Step 2. In Step 1, six annotators worked on half of the target dialogues, and the remaining six worked for other half. For collecting perceived information by ordinary people, we employed non-experts in linguistics; thus, they had different backgrounds. Their ages ranged widely from in their twenties to in their fifties. The male-to-female ratio was about 1:1.

We collected 12,723 *PerceivedInfo* instances in Step 1. The instances were grouped into 11,533 (91%) instances in Step 2. We use the grouped 11,533 instances of *PerceivedInfo* in the next step of the analysis. Detailed information on the frequency of collected *PerceivedInfo* is shown in Table 1. We collected about ten instances of *PerceivedInfo* for each utterance, which suggests that various information can be perceived from an utterance. Figure 2 shows an example of a chat-oriented dialogue and *PerceivedInfo* collected for an utterance in the dialogue.

3 Clustering of Perceived Information

To investigate what types of information constitute *PerceivedInfo*, we clustered the collected *PerceivedInfo* by using multiple working groups. Note that the clustering was done manually by multiple annotators to ensure high-quality clustering.

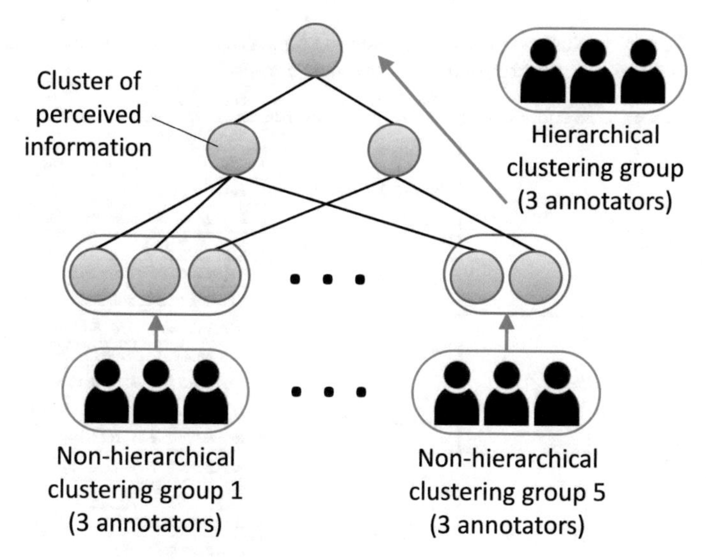

Fig. 3 Clustering procedure of perceived information

3.1 *Clustering Procedure*

The collected *PerceivedInfo* were clustered in two steps, as illustrated in Fig. 3. First, multiple working groups made disjoint clusters of *PerceivedInfo*. Second, another working group hierarchically clustered all of the created clusters. The two steps are explained below:

(Step 1) Given instances of *PerceivedInfo*, multiple working groups independently and manually clustered similar *PerceivedInfo*. Two instances of *PerceivedInfo* were regarded as similar if both indicated the same type of information. We did not provide a rigid criterion of similarity; it is decided by each working group. They labeled each cluster indicating the type of *PerceivedInfo*. They continued clustering *PerceivedInfo* until all instances of *PerceivedInfo* were contained in some cluster.

(Step 2) Another working group merged the clusters created in Step 1. They manually organized the hierarchical clusters; that is, they found the most similar groups of clusters that had similar instances of *PerceivedInfo* and merged them into a new cluster. They repeated this process until there was only one cluster. As in Step 1, they labeled each cluster indicating the type of *PerceivedInfo* contained in the cluster.

We randomly selected 300 instances of *PerceivedInfo* from five dialogues and prepared 1,500 instances of *PerceivedInfo*. For Step 1, we assigned 300 instances of *PerceivedInfo* in each dialogue for each working group consisting of three annotators. For Step 2, another group consisting of three other annotators merged the clusters.

Table 2 Hierarchical clustering results of perceived information in chat-oriented dialogue. *A* corresponds to speaker, and *B* corresponds to listener (another speaker)

First level	Second level	Third level	Fourth level
Thought (55.1%)	Belief (35.8%)	Belief self (30.7%)	Thinking (1.6%)
			The thing *A* is thinking (1.0%)
			Favorite (13.9%)
			Impression and evaluation (6.7%)
			Feeling (7.5%)
		Belief other (5.1%)	Impression of interlocutors (4.4%)
			Relation between *A* and *B* (0.7%)
	Desire (19.3%)	Desire (9.9%)	*A*'s desire related to *B* (3.1%)
			Self-contained desire (3.2%)
			A's desire 1: want to do (3.5%)
		Request (9.4%)	Wants interlocutor to do (1.1%)
			Request made to *B* (8.3%)
Fact (44.9%)	*A*'s fact (37.9%)	Attribute (20.2%)	*A*'s characteristics (19.5%)
			Possession (0.7%)
		Behavior (14.4%)	*A*'s past and experience (8.9%)
			A at present (5.5%)
		Circumstance (3.3%)	Circumstance of *A* and around *A* (1.6%)
			Circumstance around *A* (1.7%)
	Other fact (7.0%)	Certain fact (3.9%)	Fact about objects (0.7%)
			Objective fact 1: common things (0.3%)
			Objective fact (1.7%)
			Other fact: society (1.1%)
		Uncertain fact (3.1%)	Uncertain fact (1.2%)
			Fact (1.4%)
			Things that can happen (0.5%)

We recruited 15 annotators for Step 1 and three annotators for Step 2. They were different from those who collected the *PerceivedInfo*. They had different backgrounds, and two experts in linguistics. The male-to-female ratio was about 1:1.

3.2 *Types of Perceived Information*

Table 2 shows the results of *PerceivedInfo* clustering; it is hierarchical on the basis of our clustering process. The right-most column, i.e., the fourth level, corresponds to the bottom-most clusters, and their integration progresses from right to left.

From Table 2, we can see that *PerceivedInfo* is divided into the speaker's "Thought" and "Fact." "Thought" consists of speaker's "Belief" and "Desire." "Fact" consists of facts regarding a speaker, namely "*A*'s fact," and "Other fact," which is information irrelevant to the speakers. These types on the second level are further divided into the types of the third level. The details of each type on the third level are given

Table 3 Descriptions and representative examples of perceived information in third level

Belief self: Speaker's beliefs about his/herself

- Opinions: "*A* is displeased with prices in Tokyo." "*A* regards Japan as a safe country."
- Likes and Dislikes: "*A* likes playing TV games." "*A* hates smoking."
- Emotions: "*A* is excited." "*A* is happy at *B*'s praise."

Belief other: Speaker's belief toward the counterpart speaker

- Belief regarding utterances: "*A* agrees with *B*." "*A* can't believe *B*'s story."
- Belief regarding counterparts: "*A* is worried about *B*." "*A* thinks *B* is great."

Desire: Speaker's desire mainly relative to his/herself

- Desires of speakers: "*A* wants to go to Mt. Fuji." "*A* hopes summer ends soon."
- Desires relative to counterparts: "*A* wants to change the topic." "*A* wants to talk about his hobby."

Request: Speaker's requests to the counterpart

- Requests to counterparts: "*A* wants to be praised by *B*." "*A* wants to know about his hobby."
- Goals achieved with counterparts: "*A* wants to favor *B*'s opinion." "*A* wants for *B* to know what is interesting about the movie."

Attribute: User-modeling information of speakers

- Knowledge and Capability: "*A* knows a lot about cars." "*A* can drink."
- Social attributes: "*A* is woman." "*A* is married." "*A* lives in Kyoto."
- Personality: "*A* is earnest." "*A* is a determined person."

Behavior: Speaker's actions

- Habits: "*A* usually watches TV." "*A* hardly goes out." "*A* drives a car."
- Past and Experience: "*A* talked with his parents." "*A* has travelled abroad."
- State during dialogue: "*A* is trying to change the topic." "*A* seems proud." "*A* is thinking of what to say next."

Circumstance: Environment around speakers

- Relationships: "*A* is close with his parents." "*A*'s husband often watches TV."
- Living environment: "There are a lot of transfers in *A*'s job." "*A*'s parent's home is in Kanto prefecture."

Certain fact: Certain facts irrelevant to speakers

- Certain facts: "This summer is very humid." "Mt. Fuji is famous for autumn leaves."

Uncertain fact: Uncertain facts irrelevant to speakers

- Uncertain facts: "The rice crop may fail." "The economic depression is coming to an end."

in Table 3. The clustering results show that *PerceivedInfo* has various types of information including ones used in conventional studies, such as the BDI model (Belief, Desire, and Intention) [12]. It is also clear that personal information is playing an important role in dialogue.

Table 4 Inter-annotator agreement as to types of perceived information from first level to third level in clusters of perceived information

	First level	Second level	Third level
Label-wise agreement	0.90	0.86	0.75
Fleiss' κ	0.80	0.79	0.69

a1 / a2	Belief self	Belief other	Desire	Request	Attribute	Behavior	Circumstance	Certain fact	Uncertain fact
Belief self	684	63	10	1	64	50	4	2	0
Belief other	13	130	1	1	3	10	0	0	0
Desire	37	9	252	72	1	13	0	2	0
Request	2	7	14	137	0	2	0	0	0
Attribute	29	3	0	0	466	39	16	1	0
Behavior	35	24	3	1	36	499	12	5	0
Circumstance	2	0	0	0	3	2	39	1	0
Certain fact	0	0	0	0	0	1	15	75	23
Uncertain fact	1	0	0	0	0	0	3	14	68

a1 / a3	Belief self	Belief other	Desire	Request	Attribute	Behavior	Circumstance	Certain fact	Uncertain fact
Belief self	710	68	13	0	44	36	2	3	2
Belief other	10	128	0	1	0	19	0	0	0
Desire	60	19	215	77	0	14	0	0	1
Request	2	17	18	123	0	2	0	0	0
Attribute	89	6	0	0	410	29	15	4	1
Behavior	64	35	1	0	75	419	15	5	1
Circumstance	3	0	0	0	2	1	40	1	0
Certain fact	2	0	0	0	0	2	11	79	20
Uncertain fact	3	0	0	0	0	0	2	40	41

a2 / a3	Belief self	Belief other	Desire	Request	Attribute	Behavior	Circumstance	Certain fact	Uncertain fact
Belief self	682	21	24	3	34	31	4	2	2
Belief other	9	194	5	2	1	25	0	0	0
Desire	45	12	166	52	0	4	0	0	1
Request	2	18	48	144	0	0	0	0	0
Attribute	115	4	0	0	426	19	4	4	1
Behavior	76	23	4	0	60	432	14	6	1
Circumstance	9	0	0	0	8	6	60	6	0
Certain fact	2	1	0	0	2	5	3	71	16
Uncertain fact	3	0	0	0	0	0	0	43	45

	a1		a2		a3		total	
	Freq	Ratio	Freq	Ratio	Freq	Ratio	Freq	Ratio
Belief self	878	0.29	803	0.27	943	0.31	2624	0.29
Belief other	158	0.05	236	0.08	273	0.09	667	0.07
Desire	386	0.13	280	0.09	247	0.08	913	0.10
Request	162	0.05	212	0.07	201	0.07	575	0.06
Attribute	554	0.18	573	0.19	531	0.18	1658	0.18
Behavior	615	0.20	616	0.21	522	0.17	1753	0.19
Circumstance	47	0.02	89	0.03	85	0.03	221	0.02
Certain fact	114	0.04	100	0.03	132	0.04	346	0.04
Uncertain fact	86	0.03	91	0.03	66	0.02	243	0.03
total	3000	1.00	3000	1.00	3000	1.00	9000	1.00

Fig. 4 Confusion matrices and number of labels used to annotate third-level types of perceived information. Symbols from "a1" to "a3" denote each annotator

4 Evaluation

To evaluate the clusters, three annotators different from those who created the clusters annotated the labels of the *PerceivedInfo*. We used the first to third levels of clustering as annotation labels. Each annotator annotated 3,000 instances of *PerceivedInfo* that were not used in the clustering. They annotated labels by looking solely at *PerceivedInfo*, without using the context information that led to the *PerceivedInfo* in question.

Table 4 shows the inter-annotator agreement in terms of the label-wise agreement ratio and Fleiss' κ. κ values from 0.69 to 0.80 shows that there was substantial agreement between annotators. This quite high agreement indicates that the clusters cover various instances of *PerceivedInfo* and clearly distinguish each type of *PerceivedInfo*.

Figure 4 shows the confusion matrices and numbers of labels as the annotation results. The confusion matrices indicated that every pair of annotators disagreed about the "Belief self" and "Attribute" types. A representative example of this disagreement is "*A* prefers natural food." The annotators also disagreed about the "Belief self" and "Behavior" types; sometimes, belief and behavior were difficult to separate as in the case of "*A* is blushing." Our brief analysis indicates the possible need to use a combination of labels in some cases.

The table on the bottom right shows the number of labels; the distributions of each annotator's results were mostly the same as in Table 2. This result suggests that the annotation is reliable and that different dialogues may share the same distribution of *PerceivedInfo*.

5 Conclusion

We investigated the types of information that humans understand in chat-oriented dialogue. To reveal what types of information constitute the perceived information, we collected a large amount of perceived information in chat-oriented dialogue and clustered it. The types of perceived information were evaluated on the basis of inter-annotator agreement, and their validity was verified. To the best of our knowledge, this study is the first to clarify the types of information that a system would require to understand in chat-oriented dialogue systems.

In the future, we intend to develop methods for automatically extracting perceived information from dialogue and build dialogue agents that can understand users better and take appropriate actions based on the estimated perceived information. As we now have categorical perceived information, we believe that we can initiate work on estimating perceived information. Another interesting future work will be to discuss our results in terms of conventional dialogue theories, such as the cooperative principle [3], plan-based approaches to dialogue [2], and dialogue games [1].

References

1. Dialogue-games: metacommunication structures for natural language interaction. Cogn Sci 1(4), 395–420 (1977)
2. Allen J (1983) Recognizing intentions from natural language utterances. MIT Press
3. Grice HP (1975) Logic and conversation. In: Cole P, Morgan J (eds) Syntax and semantics, vol 3: Speech acts. Academic Press, pp 41–58
4. Higashinaka R, Imamura K, Meguro T, Miyazaki C, Kobayashi N, Sugiyama H, Hirano T, Makino T, Matsuo Y (2014) Towards an open domain conversational system fully based on natural language processing. In: Proceedings of COLING, pp 928–939
5. Joshi A, Sharma V, Battacharyya P (2015) Harnessing context incongruity for sarcasm detection. In: Proceedings of ACL/IJCNLP, pp 757–762

6. Kim Y, Bang J, Choi J, Ryu S, Koo S, Lee GG (2014) User information extraction for personalized dialogue systems. In: Proceedings of workshop on multimodal analyses enabling artificial agents in human-machine interaction
7. Ma C, Prendinger H, Ishizuka M (2005) Emotion estimation and reasoning based on affective textual interaction. In: Proceedings of affective computing and intelligent interaction
8. Meguro T, Minami Y, Higashinaka R, Dohsaka K (2014) Learning to control listening-oriented dialogue using partially observable markov decision processe. ACM Trans Speech Lang Process 10(15):1–15
9. Riloff E, Quadir A, Surve P, Silva LD, Gilbert N, Huang R (2013) Sarcasm as contrast between positive sentiment and negative situation. In: Proceedings of EMNLP, pp 704–714
10. Takabatake Y, Morita H, Kawahara D, Kurohashi S, Higashinaka R, Matsuo Y (2015) Classification and acquisition of contradictory event pairs using crowdsourcing. In: Proceedings of the 3rd workshop on EVENTS: definition, detection, coreference, and representation, pp 99–107
11. Williams JD, Raux A, Ramachandran D, Black, A (2013) The dialog state tracking challenge. In: Proceedings of SIGDIAL, pp 404–413
12. Wong W, Cavedon L, Thangarajah J, Padgham L (2012) Flexible conversation management using a BDI agent approach. In: Proceedings of IVA, pp 464–470
13. Zhang C, Chai JY (2009) What do we know about conversation participants: experiments on conversation entailment. In: Proceedings of SIGDIAL, pp 206–215
14. Zhang C, Chai JY (2010) Towards conversation entailment: an empirical investigation. In: Proceedings of EMNLP, pp 756–766
15. Zhang M, Qin B, Liu T, Zheng M (2014) Triple based background knowledge ranking for document enrichment. In: Proceedings of COLING, pp 917–927
16. Zhang M, Qin B, Zheng M, Hirst G, Liu T (2015) Encoding distributional semantics into triple-based knowledge ranking for document enrichment. In: Proceedings of ACL/IJCNLP, pp 524–533

Chunks in Multiparty Conversation—Building Blocks for Extended Social Talk

Emer Gilmartin, Benjamin R. Cowan, Carl Vogel and Nick Campbell

Abstract Building applications which can form a longer term social bond with a user or engage with a group of users calls for knowledge of how longer conversations work. This paper describes preliminary explorations of the structure of long (c. one hour) multiparty casual conversations, focusing on a binary distinction between two types of interaction phases—chat and chunk. A collection of long form conversations which provide the data for our explorations is described. The main result is that chat and chunk segments show differences in the distribution of their duration.

Keywords Multiparty conversation · Turntaking · Dialog systems

1 Introduction

Increasing interest in socially competent artificial spoken dialogue calls for clearer understanding of the mechanisms and form of human casual and social conversation. This knowledge can facilitate the design and implementation of applications to provide companionship, dialogic self-paced learning, entertainment and gaming, and help package information and manage interactions through natural spoken dialogue. This knowledge could also aid in machine understanding of dialogue. Dialogue technology has long focused on task-based dialogues—driven by propositional information exchange where success can be measured by efficient arrival at a clearly defined short term goal. Casual social conversation presents a problem where dialogue success does not primarily depend on the acquisition of information by one or more participants, but also on 'buy-in' or engagement in the activity of talking itself, in addition to the construction and maintenance of a social bond. In recent years, there has been significant progress in the creation of chat applications, particularly in modelling smalltalk—short casual interactions, often in the form of 'getting to know you'

E. Gilmartin (✉) · C. Vogel · N. Campbell
Trinity College Dublin, Dublin, Ireland
e-mail: gilmare@tcd.ie

B. R. Cowan
University College Dublin, Dublin, Ireland

M. Eskenazi et al. (eds.), *Advanced Social Interaction with Agents*, Lecture Notes in Electrical Engineering 510, https://doi.org/10.1007/978-3-319-92108-2_4

dialogue activities. These efforts have been supported by the creation and analysis of corpora of relatively short and often dyadic first encounter dialogues between human participants and in Wizard of Oz scenarios. Building applications which can form a longer term social bond with a user or engage with a group of users calls for knowledge of how longer conversations work. In this paper we describe preliminary explorations of the structure of long (c. one hour) multiparty casual conversations, focussing on a binary distinction between two types of interaction phases—chat and chunk. We provide a brief review of relevant existing work, describe a collection of long form conversations which provide the data for our explorations, outline some early results that may prove useful in the design of such conversations, and finally discuss our future work.

2 The Shape of Conversation—Phases, Chat, and Chunks

Talk is ubiquitous in human life. While some spoken interaction is the medium for performance of practical or instrumental tasks such as service encounters (shops, doctor's appointments), information transfer (lectures), or planning and execution of business (meetings), much daily talk serves to build and maintain social bonds, ranging from short 'bus-stop' conversations between strangers to longer sessions where friends spend time 'hanging out' engaged in what Schelgoff described as 'a continuing state of incipient talk' [13]. In these interactions, there is no clear short-term practical task or prescribed subject of discussion. Speakers are thought to have equal rights to contribute to the talk [18] or at least not to be subject to the clearly predefined roles such as 'teacher-student' which are part of task-based or instrumental encounters [4]. The form of such talk is also different to that of task-based exchanges—there is less reliance on question-answer sequences and more on commentary [16]. Instead of asking each other for information, participants seem to collaborate to fill the floor and avoid uncomfortable silence. As a simple example, a meeting has an agenda and it would be perfectly normal for the chairperson to impose the next topic for discussion. In casual conversation there is no chairperson and topics are often introduced by means of a statement or comment by a participant which may or may not be taken up by other participants.

Several researchers have noted that casual conversations develop as a sequence of phases; after initial chat where there are frequent back and forth contributions among the various participants, the structure of the talk moves to a series of longer stretches or chunks dominated by one participant at a time, interwoven with more chatty phases.

Ventola [17] described how a non-transactional conversation may comprise several structural elements or phases, which followed one another like beads on a string, sometimes repeating. The structural elements she described are:

G Greeting.
Ad Address. Defines addressee ("Hello, Mary", "Excuse me, sir")

Id Identification (of self)—only for strangers.
Ap Approach. Basically smalltalk. Can be direct (ApD)—asking about interactants themselves (so usually people who already know one another), or indirect (ApI)—talking about immediate situation (weather, surroundings, so can happen between strangers or with greater social distance). In Ventola's view, these stages allow participants to get enough knowledge about each other to enter more meaningful conversation.
C Centring. Here participants become fully involved in a conversation, talking at length. This stage is much less predictable than the Approach stage in terms of topic, and can range over several overlapping topics for an indeterminate number of repetitions, often interspersed with further Approach phases.
Lt Leave-taking. Signalling desire or need to end conversation.
Gb Goodbye. Can be short or extended, in which case there are projections to further meetings.

Ventola develops a number of sequences of these elements for conversations involving different levels of social distance. She describes conversations as minimal or non-minimal, where a minimal conversation is essentially phatic, particularly in Jakobsen's sense of maintaining channels of communication [10], or Schneider's [14] notion of defensive smalltalk—such a conversation could simply be a greeting, or could be a chatty sequence of approach stages. Non-minimal conversations involve centring—where the focus shifts to longer bouts often fixed on a particular topic. Several of the elements are optional and omitted in particular situations. For example, friends can jump from Greeting to Centring without passing through the 'smalltalk' exploratory Approach stages. Strangers may not greet one another but could start with ApI ("It's a nice day"). Many elements occur only once, such as greetings (G) and goodbyes (Gb), but others can recur. Approach stages can occur recursively, generating long chats without getting any deeper into centring. Centring stages can recur and are often interspersed with Approach stages in longer talks. Figure 1 shows a simplified version of Ventola's model.

Another view of the structure of causal conversation has been developed by Slade and Eggins, who regard casual talk as sequences of 'chat' and 'chunk' elements [7]. Chat segments are highly interactive and appear to be managed locally, unfolding move by move or turn by turn, and are thus amenable to Conversation Analysis style study. Chunks are segments where (i) 'one speaker takes the floor and is allowed to dominate the conversation for an extended period', and (ii) the chunk appears to move through predictable stages, amenable to genre analysis. In a study of three hours of conversational data collected during coffee breaks in three different workplaces involving all-male, all-female, and mixed groups, Slade found that around fifty percent of all talk could be classified as chat, while the rest comprised longer form chunks from the following genres: storytelling, observation/comment, opinion, gossip, joke-telling and ridicule.

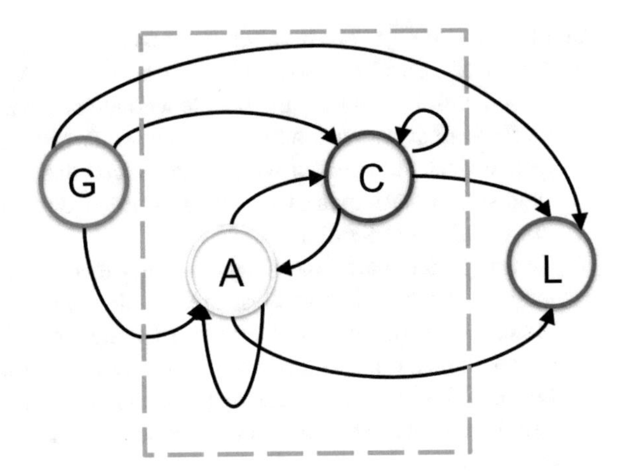

Fig. 1 A simplified version of Ventola's conversational phases—Greeting (G) and Leavetaking (L) occur at most once in a conversation, while the Approach (A) and Centring (C) phases may repeat and alternate indefinitely in a longer conversation. Different conversational sequences may be generated from the graph

Ventola's phases and Slade and Eggin's binary distinctions (and more detailed generic classification of chunks) could greatly aid the segmentation of conversations into phases or subroutines, which could be used in the design and management of artificial dialogues.

There has been work on the theory and analysis of aspects of social conversation, often covering particular phases, such as the long tradition of studies of the structure of narrative, recently applied to the Switchboard spoken corpus [5], or to the patterning of speaker turns in narratives from spoken dialogues in the British National Corpus [12]. There has also been work on creating smalltalk dialogues as part of more task-based talk [1], or alone [20], and on understanding the development of relationships between interlocutors [6, 15]. Much of this work has focused on dyadic exchanges. In the preliminary explorations described below, we focus on multiparty conversation, and take a broad view of chat and chunks rather than drilling down to specific genres within chunks.

3 Data and Binary Annotation of Chat and Chunk Phases

To aid our understanding of the conversational phases in our data, we annotated six long form conversations for chat and chunk. We then extracted descriptive statistics which we report below. We have also marked up the conversations using Ventola's phases which will form the basis for further studies.

The six conversations were drawn from three multimodal corpora of multiparty casual talk—d64, DANS, and TableTalk. In all three corpora there were no instructions to participants about what to talk about and care was taken to ensure that all participants understood that they were free to talk or not as the mood took them. The d64 corpus is a multimodal corpus of informal conversational English recorded in Dublin in 2009 in an apartment living room with 2–5 people on camera at all times

Table 1 Dataset for experiments

Conversation	Corpus	Participants	Gender	Duration (s)
A	D64	5	2F/3M	4164
B	DANS	3	1F/2M	4672
C	DANS	4	1F/3M	4378
D	DANS	3	2F/1M	3004
E	TableTalk	4	2F/2M	2072
F	TableTalk	5	3F/2M	4740

[11]. The DANS corpus contains conversations ranging between 60-90 min with 2 to 4 participants in a living-room setup in the Speech Communication Lab in Dublin in 2012 [9]. The TableTalk corpus was recorded at the Advanced Telecommunications Research Institute (ATR) in Kyoto in 2007, and consists of 3 sessions of 4 or 5-party casual conversations of around 90 min in duration [3]. Details of the conversations are shown in Table 1.

Frequent overlap and bleedover from other speakers in the audio recordings made them unsuitable for automatic segmentation, so the conversations were manually segmented into speech (including laughter) and silence using Praat [2] and Elan [19]. The segmentation, transcription and annotation of the data are more fully described in [8].

For an initial classification, conversations were segmented into phases by first identifying all of the 'chunks' using the first, structural part of Slade & Eggins' definition—'a segment where one speaker takes the floor and is allowed to dominate the conversation for an extended period' [7]. All other interaction was considered chat.

4 Experiments

The annotations resulted in 213 chat segments and 358 chunk segments overall. Preliminary inspection of the data showed that the distributions were unimodal but heavily right skewed, with a hard left boundary at 0. Log durations were closer to normal with skew reduced from 1.621 to 0.004 for chat and from 1.935 to 0.237 for chunks. These values are below the generally accepted 0.5 threshold for near normality. It was decided to use geometric means to describe central tendencies in the data, after removing one outlying value in the log durations(>1.5 times IQR).

However, it should be noted that several conversations shared speakers and the number of chunk segments produced by speakers varied widely (8:68), and thus we decided to create a balanced sample to minimize bias. We took the entire sample of the speaker with the fewest chunks (8) and created random samples ($n = 8$) of chunks for each of the other speakers. This resulted in a sample of 96 chunks.

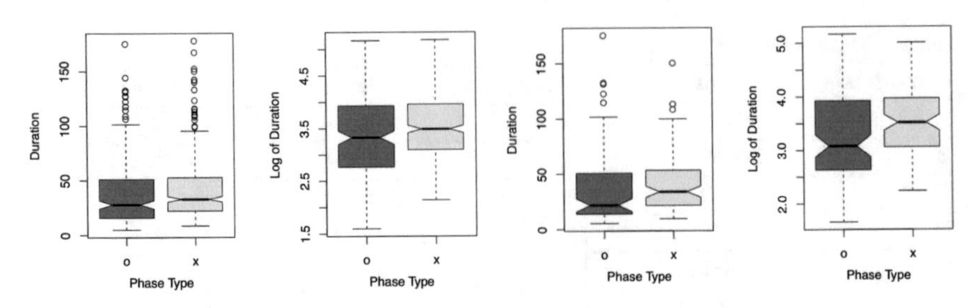

Fig. 2 Boxplots of distributions of duration and log durations of chat ('o') and chunk ('x') phases in entire dataset (left) and balanced sample (right)

We also extracted a random sample of 96 chat segments from the data for comparison purposes. The log transform here reduced skew from 1.772 to 0.236 for chat and from 1.523 to 0.015 for chunks.

Figure 2 shows the boxplots of raw and log durations for chat and chunk segments in both the full dataset and the balanced sample.

The antilogs of geometric means for duration of chat and chunk phases in the original dataset were 28.1 seconds for chat and 34 seconds for chunks, while in the balanced sample the chat value was 25.2 and the chunk value was 33.2. These values were close to the median in all cases, in contrast to the elevated mean values seen for the untransformed data.

Mann-Whitney U tests on both the full data set and the balanced subset showed significant differences in the distributions of the untransformed durations for chat and chunk ($p < 0.01$ for the full set, $p < 0.05$ for the smaller balanced sample).

5 Conclusions and Future Work

The first result of interest in our preliminary explorations is that there is a difference in the distributions of chat and chunk durations—chat varies more while chunks have a stronger central tendency. This could indicate that there is a natural limit for the time one speaker should dominate a conversation and this knowledge could be used in system design. The mean duration of chunk phases was consistent between the full dataset and the balanced sample, which may indicate that chunk duration is not speaker dependent. The 'half a minute' mean duration of chunks could prove useful in the design of pedagogical dialogues, allowing information to be 'dosed' in natural chunks which could be easier for learners to assimilate. The larger number of chunk phases in the data compared to Slade's findings on work break conversations may be due to the length of the conversations examined here—we found several instances of sequential chunks where the long turn passed directly to another speaker without intervening chat. Systems which understand and/or generate social human-machine interaction need ground truths based on relevant data in order to create accurate models. We hope that our further explorations into the architecture of longer form conversation will add to this body of knowledge. In addition to studying multiparty data, we intend to investigate dyadic conversations. Unfortunately, there is a dearth of long form conversational data avaliable for analysis. Hopefully, the proven value of past task-based data corpora and the growing importance of social human-machine spoken dialogue will encourage the collection of larger datasets of casual or social conversation open to the research community.

Acknowledgements This work is supported by the European Coordinated Research on Long-term Challenges in Information and Communication Sciences and Technologies ERA-NET (CHIS-TERA) JOKER project, JOKe and Empathy of a Robot/ECA: Towards social and affective relations with a robot, and by the Speech Communication Lab, Trinity College Dublin.

References

1. Bickmore T, Cassell J (2005) Social dialongue with embodied conversational agents. Adv Nat Multimodal Dialogue Syst, 23–54
2. Boersma P, Weenink D (2010) Praat: doing phonetics by computer [Computer program]. Version 5(1):44
3. Campbell N (2008) Multimodal processing of discourse information: the effect of synchrony. In: Second international symposium on universal communication, ISUC'08, pp 12–15
4. Cheepen C (1988) The predictability of informal conversation. Pinter London
5. Collins KJ, Traum D (2016) Towards a multi-dimensional taxonomy of stories in dialogue. In: Chair NCC, Choukri K, Declerck T, Goggi S, Grobelnik M, Maegaard B, Mariani J, Mazo H, Moreno A, Odijk J, Piperidis S (eds) Proceedings of the tenth international conference on language resources and evaluation (LREC 2016). European Language Resources Association (ELRA), Paris, France
6. Devillers L, Rosset S, Duplessis GD, Sehili MA, Béchade L, Delaborde A, Gossart C, Letard V, Yang F, Yemez Y, Türker BB, Sezgin M, Haddad KE, Dupont S, Luzzati D, Esteve Y, Gilmartin E, Campbell N (2015) Multimodal data collection of human-robot humorous interactions in the joker project. In: 2015 international conference on affective computing and intelligent interaction (ACII), pp 348–354
7. Eggins S, Slade D (2004) Analysing casual conversation. Equinox Publishing Ltd
8. Gilmartin E, Campbell N (2016) Capturing chat: annotation and tools for multiparty casual conversation. In: Proceedings of the tenth international conference on language resources and evaluation (LREC 2016)
9. Hennig S, Chellali R, Campbell N (2014) The D-ANS corpus: the Dublin-Autonomous Nervous System corpus of biosignal and multimodal recordings of conversational speech. Reykjavik, Iceland
10. Jakobson R (1960) Closing statement: linguistics and poetics. Style Lang 350:377
11. Oertel C, Cummins F, Edlund J, Wagner P, Campbell N (2010) D64: a corpus of richly recorded conversational interaction. J Multimodal User Interfac, 1–10 (2010)
12. Ruhlemann C, Gries S (2015) Turn order and turn distribution in multi-party storytelling. J Pragmat 87
13. Schegloff E, Sacks H (1973) Opening up closings. Semiotica 8(4):289–327
14. Schneider KP (1988) Small talk: analysing phatic discourse, vol 1. Hitzeroth Marburg
15. Schulman D, Bickmore T (2010) Modeling behavioral manifestations of coordination and rapport over multiple conversations. In: Intelligent virtual agents, pp 132–138
16. Thornbury S, Slade D (2006) Conversation: from description to pedagogy. Cambridge University Press
17. Ventola E (1979) The structure of casual conversation in English. J Pragmat 3(3):267–298

18. Wilson J (1989) On the boundaries of conversation, vol 10. Pergamon
19. Wittenburg P, Brugman H, Russel A, Klassmann A, Sloetjes H (2006) Elan: a professional framework for multimodality research. In: Proceedings of LREC, vol 2006
20. Yu Z, Xu Z, Black AW, Rudnicky A (2016) Strategy and policy learning for non-task-oriented conversational systems. In: Proceedings of the 17th annual meeting of the special interest group on discourse and dialogue

Debbie, the Debate Bot of the Future

Geetanjali Rakshit, Kevin K. Bowden, Lena Reed, Amita Misra and Marilyn Walker

Abstract Chatbots are a rapidly expanding application of dialogue systems with companies switching to bot services for customer support, and new applications for users interested in casual conversation. One style of casual conversation is argument; many people love nothing more than a good argument. Moreover, there are a number of existing corpora of argumentative dialogues, annotated for agreement and disagreement, stance, sarcasm and argument quality. This paper introduces Debbie, a novel arguing bot, that selects arguments from conversational corpora, and aims to use them appropriately in context. We present an initial working prototype of Debbie, with some preliminary evaluation and describe future work.

Keywords Debate · Chatbot · Argumentation · Conversational agents

1 Introduction

A chatbot or a conversational agent is a computer program that can converse with humans, via speech or text, with the goal of conducting a natural conversation, hopefully indistinguishable from a real human-to-human interaction. Chatbots are gaining momentum as more companies are switching to bot services for customer care, but there is also an opportunity for different types of casual conversation.

G. Rakshit (✉) · K. K. Bowden · L. Reed · A. Misra · M. Walker
Natural Language and Dialogue Systems Lab, University of California, Santa Cruz, USA
e-mail: grakshit@ucsc.edu

K. K. Bowden
e-mail: kkbowden@ucsc.edu

L. Reed
e-mail: lireed@ucsc.edu

A. Misra
e-mail: amisra2@ucsc.edu

M. Walker
e-mail: mawalker@ucsc.edu

M. Eskenazi et al. (eds.), *Advanced Social Interaction with Agents*, Lecture Notes in Electrical Engineering 510, https://doi.org/10.1007/978-3-319-92108-2_5

Conversational agents can be broadly classified into retrieval-based and generative models. Retrieval-based methods have a repository of predefined responses and a mechanism to pick an appropriate response based on user input. They, therefore, can't generate a completely new response. Such methods are commonly used for "help system" chatbots, that target a predefined set of FAQs and responses. Another strategy is to use rule-based expression matching, and templated response generation, as in ELIZA or JULIA [10, 20, 34]. Most existing chatbots are task-oriented and their evaluation is based on the successful accomplishment of the task.

There are many websites dedicated to debating controversial topics, and data from them have been structured and labeled in previous work. For example, we have access to trained models for labeling these argumentative conversations with attributes such as agreement, disagreement, stance, sarcasm, factual versus feeling arguments, argument quality and argument facets. This data provides us with a rich source of conversational data for mining argumentative responses. We build on previous work in our lab on disagreement detection, classifying stance, identifying high quality arguments, measuring the properties and the persuasive effects of factual versus emotional arguments, and clustering arguments into their facets or frames related to a particular topic [1, 18, 22, 23, 25, 30, 31].

In this work, we present Debbie, a novel arguing bot, that uses retrieval from existing conversations in order to argue with users. Debbie's main aim is to keep the conversation going, by successfully producing arguments and counter-arguments that will keep the user talking about the topic. Our initial prototype of Debbie works with three topics: death penalty, gun control, gay marriage. This paper focuses on our basic investigations on the initial prototype. While we are aware of other retrieval based chatbot systems [2, 4, 8, 24], Debbie is novel in that it is the first to deal with argument retrieval.

2 Data

Social media conversations are a good source of argumentative data but many sentences either do not express an argument or cannot be understood out of context and hence cannot be used to build Debbie's response pool. Swanson et al. in [30] created a large corpus consisting of 109,074 posts on the topics gay marriage (GM, 22425 posts), gun control (GC, 38102 posts), death penalty (DP, 5283 posts) by combining the Internet Argument Corpus(IAC) [32], with dialogues from online debate forums[1] [30]. It includes topic annotations, response characterizations (4forums), and stance. They build an argument quality regressor to rate the **argument quality (AQ)** using a continuous slider ranging from hard (0.0) to easy to interpret (1.0). The AQ score is intended to reflect how easily the speaker's argument can be understood from the sentence without any context. Easily understandable sentences are assumed to be prime candidates for Debbie's response pool. Misra et al. in [22] note that a thresh-

[1] http://www.createdebate.com/.

old of predicted AQ >0.55 maintained both diversity and quality in the arguments. For example, the sentence *The death penalty is also discriminatory in its application what i mean is that through out the world the death penalty is disproportionately used against disadvantaged people* was given a score of 0.98.

We started with the Argument Quality (AQ) regressor from [30], which predicts a quality score for each sentence. The stance for these argument segments is obtained from IAC [1]. We keep only stance bearing statements from the above dataset. Misra et al. in [22] had improved upon the AQ predictor from [30], giving a much larger and diverse corpus. Since generating a cohesive dialogue is a challenging task, we first evaluated our prototype with hand labeled 2000 argument quality sentence pairs for the topic of death penalty obtained from [22]. We tested our model for both appropriateness of responses and response times. Once we had a working system for death penalty, we added the best quality 250 arguments for gay marriage and gun control, each, from the corpus of [22] (This had 174405 arguments from gay marriage and 258763 for gun control).

3 Methodology

Debbie has domain knowledge of three hot button topics - death penalty, gay marriage and gun control. We created a database of statements for and against each topic, which serves as the source for Debbie's views.

The user picks a topic from a pool of topics and specifies his/her stance (for or against). As the user provides an argument, Debbie uses a similarity algorithm to retrieve a ranked list of the most appropriate counter-arguments, i.e., arguments opposing the user's stance. To speed up this retrieval process, we pre-create clusters of the arguments present in our database (described in Sect. 3). The most appropriate counter-arguments are calculated based on a similarity score, which, in this case, was the UMBC STS score [14]. The similarity algorithm takes as input two sentences and returns a real-valued score, which we use directly as the similarity between two argument statements. It uses a lexical similarity feature that combines LSA (Latent Semantic Similarity) word similarity, and WordNet knowledge, and can be run from a web API provided by the authors [14].

Debbie's responses are stored for the duration of the chat. From the ranked list, the most appropriate response (having the highest similarity score), that has not already been used in the chat, is selected. Since we only have high quality arguments in our database, we expect Debbie's responses to be good in terms of grammatical correctness, on topic, etc. Another way of looking at appropriateness of a response is basically how adequate a retort it is to the user's view. Debbie sustains the debate until the user explicitly ends the chat. While there is a limited number of unique counter-arguments which Debbie can utilize, it would take the user substantial time to exhaust all possible responses.

3.1 Generating Clusters

To create the clusters, we first generated a distance matrix of similarity scores for each topic and stance. A similarity score falls between 0 and 1. Using agglomerative clustering from scikit-learn, we created 15 clusters. We then identified the head of a cluster; the argument within each cluster, that best represents all of the statements within that cluster. We calculated this by finding the average distance for each statement in a cluster to all the statements in the cluster, and chose the one with the minimum average as the head. Hierarchical agglomerative clustering has been previously used for argument clustering by [5].

3.2 Using the Clusters

To speed up the response times by clustering, we compare the user's input to the head of each cluster. We find the cluster whose head has the highest similarity score and calculate the similarity score of each response within that cluster in order to return the most similar response.

We further optimized our algorithm by implementing a graph-based comparison method to find an acceptable cluster faster. We create a graph with the cluster heads as nodes. We start at a random head and find how similar it is to the input. If the similarity passes a high threshold of 0.9, we use the related cluster automatically. Otherwise, if the similarity is very high, say, above a threshold of 0.8, we eliminate connected edges where the similarity is very low, below a threshold of 0.5. Conversely, if the similarity is very low, say, below 0.5, we eliminate connected edges where the similarity is very high, above 0.8. In the case where we don't find a head which surpasses our high threshold, we continue to explore our graph until all the clusters have either been visited or eliminated from consideration. We then select the head with the highest similarity score.

4 Evaluation

A sample conversation with Debbie is shown in Fig. 1, where the user supports the death penalty and Debbie opposes it. For a start, we looked at methods for faster response retrieval and the quality of the responses. The most basic (baseline) method just finds the similarity score between the input and each possible response in our database, and returns the response with the highest similarity score. The second method is the simple clustering method and the third method is clustering with the graph method described in Sect. 3.

Table 1 shows the average response times for each retrieval method. We arrived at these by testing for three sentences per stance per topic, each deliberately chosen

Fig. 1 Chat where Debbie is against the death penalty

User: I think some criminals just deserve to die.
Debbie: He has his reasons that you and I can't even begin to comprehend, who are we kidding when we think we can say who deserves to live and die.
User: Why should taxpayers pay to keep criminals alive?
Debbie: And sure it is at a cost to the taxpayer, but whether or not we execute someone should not be dependent upon the relative financial cost.
User: Some people feel no remorse for their crimes anyway, nothing good can come out of keeping them alive.
Debbie: When they are about to be shot just think what would be happening, some unlucky person would be dealing with the fact that they were going to have to take someone's life.

Table 1 Average response times in seconds

Topic	Stance	Baseline	Cluster	Graph
DP	for	60.6	3.9	8.1
DP	against	55.5	7.9	5.0
GC	for	70.7	25.2	24.3
GC	against	73.5	22.9	15.3
GM	for	62.8	10.0	9.2
GM	against	62.8	3.2	2.9

such that they would access different clusters. As expected, both the cluster and the cluster graph methods perform faster than the baseline. The cluster graph method is faster than using just clusters in most cases. The exception was the "for" case of death penalty, which, we believe, can be attributed to the significantly larger size of the cluster that is accessed by the cluster graph, triggering a greater number of computations. Given that our database only has high quality arguments, the appropriateness of Debbie's responses are primarily dependent on the performance of the similarity algorithm. However, exchange of only high quality arguments between the system and the user, with minimal repetition (none from Debbie) hampers the natural flow of the conversation.

5 Future Work

The prototype we are proposing represents our pilot work with Debbie, an argumentative chatbot. Our work in this domain is still in progress and we have a lot of future work planned based on our preliminary observations. We acknowledge the fact that we must migrate Debbie away from the assumption that the argument as a whole will consist of argumentatively sound statements. In our evaluation, we observed a high usage of utterances such as *You're just wrong.* and *I don't think so.* from the user. These statements have low argumentative quality, or, are completely off-topic. Hence, responding to them with a high quality argument tended to sound unnatural and inappropriate. Therefore, in order to make the conversation sound less robotic and more natural, we must detect user utterances which are not argumentatively sound and respond accordingly.

Lukin et al. [18] talk about the role of personality in persuasion and Bowden et al. [6] have shown that adding a layer of stylistic variation to a dialogue is sufficient for representing personality between speakers. We intend to investigate how we can enhance the user's experience by entraining Debbie's personality with respect to the user's personality. While our initial results are promising, we need to improve Debbie's performance with regards to retrieval time. We can potentially do this by recursively employing the graph method within the clusters - similar to hierarchical clusters. We also intend to explore alternative information retrieval methods such as indexing, to create a more balanced trade-off between appropriateness and response time. Debbie currently uses the UMBC STS for calculating similarity scores, which is known to not work very well for argumentative sentences [22]. We intend to use a better argument similarity algorithm in the future.

References

1. Abbott R, Ecker B, Anand P, Walker, M (2016) Internet argument corpus 2.0: an SQL schema for dialogic social media and the Corpora to go with it. LREC
2. Ameixa D, Coheur L, Fialho P, Quaresma P (2014) Luke, i am your father: dealing with out-of-domain requests by using movies subtitles. In: International conference on intelligent virtual agents, Springer, Cham, pp 13–21
3. Andrews P, De Boni M, Manandhar S, De M (2016) Persuasive argumentation in human computer dialogue. In AAAI spring symposium: argumentation for consumers of healthcare, pp 8–13
4. Banchs RE, Li H (2012) IRIS: a chat-oriented dialogue system based on the vector space model. In: Proceedings of the ACL 2012 system demonstrations, pp 37–42. Association for Computational Linguistics
5. Boltui F, Snajder J (2015) Identifying prominent arguments in online debates using semantic textual similarity. In: Proceedings of the 2nd workshop on argumentation mining, pp 110–115
6. Bowden KK, Lin GI, Reed LI, Tree JEF, Walker MA (2016) M2d: monolog to dialog generation for conversational story telling. In: International conference on interactive digital storytelling. Springer, Cham, pp 12–24
7. Conrad A, Wiebe J (2012) Recognizing arguing subjectivity and argument tags. In Proceedings of the workshop on extra-propositional aspects of meaning in computational linguistics, pp 80–88. Association for Computational Linguistics
8. Duplessis GD, Letard V, Ligozat AL, Rosset S (2016) Purely corpus-based automatic conversation authoring. In: 10th edition of the language resources and evaluation conference (LREC)
9. Elhadad Michael (1995) Using argumentation in text generation. J Pagmatics 24(1–2):189–220
10. Foner LN (1997) Entertaining agents: a sociological case study. In: Proceedings of the first international conference on autonomous agents. ACM, pp 122–129
11. Goudas T, Louizos C, Petasis G, Karkaletsis V (2014) Argument extraction from news, blogs, and social media. In: Hellenic conference on artificial intelligence. Springer, Cham, pp 287–299

12. Habernal I, Gurevych I (2015) Exploiting debate portals for semi-supervised argumentation mining in user-generated web discourse. In: Proceedings of the 2015 conference on empirical methods in natural language processing, pp 2127–2137
13. Habernal I, Gurevych I (2016) Which argument is more convincing? analyzing and predicting convincingness of web arguments using bidirectional lstm. In: Proceedings of the 54th annual meeting of the association for computational linguistics (Volume 1: Long Papers) Vol 1, pp 1589–1599
14. Han L, Kashyap A, Finin T, Mayfield J, Weese J (2013) UMBC_EBIQUITY-CORE: semantic textual similarity systems. second joint conference on lexical and computational semantics (* SEM), Volume 1: proceedings of the main conference and the shared task: semantic textual similarity, Vol 1
15. Hasan KS, Ng V (2014) Why are you taking this stance? identifying and classifying reasons in ideological debates. In: Proceedings of the 2014 conference on empirical methods in natural language processing (EMNLP), pp 751–762
16. Higashinaka R, Imamura K, Meguro T, Miyazaki C, Kobayashi N, Sugiyama H, Matsuo Y (2014) Towards an open-domain conversational system fully based on natural language processing. In: Proceedings of COLING 2014, the 25th international conference on computational linguistics: technical papers, pp 928–939
17. Liu C-W, Lowe R, Serban IV, Noseworthy M, Charlin L, Pineau J (2016) ZIB structure prediction pipeline: how NOT to evaluate your dialogue system: an empirical study of unsupervised evaluation metrics for dialogue response generation. arXiv preprint arXiv:1603.08023
18. Lukin S, Anand P, Walker M, Whittaker S, Argument strength is in the eye of the beholder: audience effects in persuasion
19. Al-Khatib K, Wachsmuth H, Hagen M, Khler J, Stein B (2016) Cross-domain mining of argumentative text through distant supervision. In: Proceedings of the 2016 conference of the north american chapter of the association for computational linguistics: human language technologies, pp 1395–1404
20. Mauldin ML (1994) Chatterbots, tinymuds, and the turing test: entering the loebner prize competition. AAAI 94:16–21
21. Misra A, Anand P, Tree JF, Walker M (2015) Using summarization to discover argument facets in online idealogical dialog. NAACL HLT 2015
22. Misra A, Ecker B, Walker MA (2016) Measuring the similarity of sentential arguments in dialog. In: 17th annual meeting of the special interest group on discourse and dialogue, p 276
23. Misra A, Walker MA (2015) Topic independent identification of agreement and disagreement in social media dialogue. Evolution 460(460):920
24. Nio L, Sakti S, Neubig G, Toda T, Adriani M, Nakamura S (2014) Developing non-goal dialog system based on examples of drama television. In: Natural interaction with robots, knowbots and smartphones, Springer, New York, NY, pp 355–361
25. Oraby S, Reed L, Compton R, Riloff E, Walker M, Whittaker S, That AF (2015) Distinguishing factual and emotional argumentation in online dialogue. NAACL HLT 2015, 116
26. Oraby S, Harrison V, Reed L, Hernandez E, Riloff E, Walker M (2016) Creating and characterizing a diverse corpus of sarcasm in dialogue. In: 17th annual meeting of the special interest group on discourse and dialogue, p 31
27. Persing I, Ng V (2015) Modeling argument strength in student essays. In: Proceedings of the 53rd annual meeting of the association for computational linguistics and the 7th international joint conference on natural language processing (Volume 1: Long Papers). Vol 1, pp 543–552
28. Rinott R, Dankin L, Perez CA, Khapra MM, Aharoni E, Slonim N (2015) Show me your evidence-an automatic method for context dependent evidence detection. In: Proceedings of the 2015 conference on empirical methods in natural language processing, pp 440–450
29. Stab C, Kirschner C, Eckle-Kohler J, Gurevych I (2014) Argumentation mining in persuasive essays and scientific articles from the discourse structure perspective. In: ArgNLP, pp 21–25
30. Swanson R, Ecker B, Walker M (2015) Argument mining: extracting arguments from online dialogue. In: Proceedings of the 16th annual meeting of the special interest group on discourse and dialogue, pp 217–226

31. Walker MA, Anand P, Abbott R, Grant R (2012) Stance classification using dialogic properties of persuasion. In: Proceedings of the 2012 conference of the north american chapter of the association for computational linguistics: human language technologies, pp 592–596. Association for Computational Linguistics
32. Walker MA, Anand P, Abbott R, Tree JEF, Martell C, King J (2012) That is your evidence?: classifying stance in online political debate. Decis Support Syst 53(4):719–729
33. Wang L, Ling W (2016) Neural network-based abstract generation for opinions and arguments. arXiv preprint arXiv:1606.02785
34. Weizenbaum J (2966) ELIZAa computer program for the study of natural language communication between man and machine. Commun ACM 9(1): 36–45
35. Zukerman I, McConachy R, Korb, KB (2000) Using argumentation strategies in automated argument generation. Proceedings of the first international conference on Natural language generation, Vol 14. Association for Computational Linguistics, pp 55–62

Data-Driven Dialogue Systems for Social Agents

Kevin K. Bowden, Shereen Oraby, Amita Misra, Jiaqi Wu, Stephanie Lukin and Marilyn Walker

Abstract In order to build dialogue systems to tackle the ambitious task of holding social conversations, we argue that we need a data-driven approach that includes insight into human conversational "chit-chat", and which incorporates different natural language processing modules. Our strategy is to analyze and index large corpora of social media data, including Twitter conversations, online debates, dialogues between friends, and blog posts, and then to couple this data retrieval with modules that perform tasks such as sentiment and style analysis, topic modeling, and summarization. We aim for personal assistants that can learn more nuanced human language, and to grow from task-oriented agents to more personable social bots.

Keywords Dialogue systems · Data-driven approaches · Chatbots

1 From Task-Oriented Agents to Social Bots

Devices like the Amazon Echo and Google Home have entered our homes to perform task-oriented functions, such as looking up today's headlines and setting reminders [1, 5]. As these devices evolve, we have begun to expect social conversation, where the device must learn to personalize and produce natural language style.

K. K. Bowden (✉) · S. Oraby · A. Misra · J. Wu · S. Lukin · M. Walker
Natural Language and Dialogue Systems Lab, University of California, Santa Cruz, Oakland, CA, USA
e-mail: kkbowden@ucsc.edu

S. Oraby
e-mail: soraby@ucsc.edu

A. Misra
e-mail: amisra2@ucsc.edu

J. Wu
e-mail: jwu64@ucsc.edu

S. Lukin
e-mail: slukin@ucsc.edu

M. Walker
e-mail: mawalker@ucsc.edu

M. Eskenazi et al. (eds.), *Advanced Social Interaction with Agents*, Lecture Notes in Electrical Engineering 510, https://doi.org/10.1007/978-3-319-92108-2_6

Social conversation is not explicitly goal-driven in the same way as task-oriented dialogue. Many dialogue systems in both the written and spoken medium have been developed for task-oriented agents with an explicit goal of restaurant information retrieval, booking a flight, diagnosing an IT issue, or providing automotive customer support [6, 8, 18, 23, 25, 26]. These tasks often revolve around question answering, with little "chit-chat". Templates are often used for generation and state tracking, but since they are optimized for the task at hand, the conversation can either become stale, or maintaining a conversation requires the intractable task of manually authoring many different social interactions that can be used in a particular context.

We argue that a social agent should be spontaneous, and allow for human-friendly conversations that do not follow a perfectly-defined trajectory. In order to build such a conversational dialogue system, we exploit the abundance of human-human social media conversations, and develop methods informed by natural language processing modules that model, analyze, and generate utterances that better suit the context.

2 Data-Driven Models of Human Language

A myriad of social media data has led to the development of new techniques for language understanding from open domain conversations, and many corpora are available for building data-driven dialogue systems [19, 20]. While there are differences between how people speak in person and in an online text-based environment, the social agents we build should not be limited in their language; they should be exposed to many different styles and vocabularies. Online conversations can be repurposed in new dialogues, but only if they can be properly indexed or adapted to the context. Data retrieval algorithms have been successfully employed to co-construct an unfolding narrative between the user and computer [22], and re-use existing conversations [7]. Other approaches train on such conversations to analyze sequence and word patterns, but lack detailed annotations and analysis, such as emotion and humor [10, 21, 24]. The large Ubuntu Dialogue Corpus [12] with over 7 million utterances is large enough to train neural network models [9, 11].

We argue that combining data-driven retrieval with modules for sentiment analysis and style, topic analysis, summarization, paraphrasing, rephrasing, and search will allow for more human-like social conversation [3]. This requires that data be indexed based on domain and requirement, and then retrieve candidate utterances based on dialogue state and context. Likewise, in order to avoid stale and repetitive utterances, we can alter and repurpose the candidate utterances; for example, we can use paraphrase or summarization to create new ways of saying the same thing, or to select utterance candidates according to the desired sentiment [14, 15]. The style of an utterance can be altered based on requirements; introducing elements of sarcasm, or aspects of factual and emotional argumentation styles [16, 17]. Changes in the perceived speaker personality can also make more personable conversations [13]. Even utterances from monologic texts can be leveraged by converting the content to dialogic flow, and performing stylistic transformations [2].

Of course, while many data sources may be of interest for indexing knowledge for a dialogue system, annotations are not always available or easy to obtain. By using machine learning models designed to classify different classes of interest, such as sentiment, sarcasm, and topic, data can be bootstrapped to greatly increase the amount of data available for indexing and utterance selection [17].

There is no shortage of human generated dialogues, but the challenge is to analyze and harness them appropriately for social-dialogue generation. We aim to combine data-driven methods to repurpose existing social media dialogues, in addition to a suite of tools for sentiment analysis, topic identification, summarization, paraphrase, and rephrasing, to develop a socially-adept agent that can carry on a natural conversation [4].

References

1. Amazon Echo. https://www.amazon.com/Amazon-Echo-Bluetooth-Speaker-with-WiFi-Alexa/dp/B00X4WHP5E
2. Bowden KK, Lin GI, Reed LI, Tree JEF, Walker MA (2016) M2D: monolog to dialog generation for conversational story telling. In: Interactive storytelling—9th international conference on interactive digital storytelling, ICIDS 2016, Los Angeles, CA, USA, November 15–18, 2016, Proceedings, pp 12–24
3. Bowden KK, Oraby S, Wu J, Misra A, Walker M (2017) Combining search with structured data to create a more engaging user experience in open domain dialogue. SCAI 2017
4. Bowden KK, Wu J, Oraby S, Misra A, Walker M (2017) Slugbot: an application of a novel and scalable open domain socialbot framework. In: Alexa prize proceedings
5. Google Home Features. https://madeby.google.com/home/features/
6. Henderson M, Thomson B, Williams J (2014) The second dialog state tracking challenge. In: Proceedings of SIGDIAL. ACL association for computational linguistics. https://www.microsoft.com/en-us/research/publication/the-second-dialog-state-tracking-challenge/
7. Higashinaka R, Imamura K, Meguro T, Miyazaki C, Kobayashi N, Sugiyama H, Hirano T, Makino T, Matsuo Y (2014) Towards an open-domain conversational system fully based on natural language processing. In: COLING. ACL
8. Hirschman L (2000) Evaluating spoken language interaction: experiences from the darpa spoken language program 1990–1995. MIT Press, Cambridge, Mass, Spoken Language Discourse
9. Kadlec R, Schmid M, Kleindienst J (2015) Improved deep learning baselines for ubuntu corpus dialogs. CoRR **abs/1510.03753**. http://arxiv.org/abs/1510.03753
10. Li J, Galley M, Brockett C, Gao J, Dolan B (2016) A persona-based neural conversation model. arXiv preprint http://arxiv.org/abs/1603.06155
11. Lowe RT, Pow N, Serban IV, Charlin L, Liu C, Pineau J (2017) Training end-to-end dialogue systems with the ubuntu dialogue corpus. D&D 8(1): 31–65. http://dad.uni-bielefeld.de/index.php/dad/article/view/3698
12. Lowe R, Pow N, Serban I, Pineau J (2015) The ubuntu dialogue corpus: a large dataset for research in unstructured multi-turn dialogue systems. In: Proceedings of the SIGDIAL 2015 conference, the 16th annual meeting of the special interest group on discourse and dialogue, 2–4 September 2015, Prague, Czech Republic, pp 285–294. http://aclweb.org/anthology/W/W15/W15-4640.pdf
13. Lukin S, Ryan JO, Walker MA (2014) Automating direct speech variations in stories and games. In: Proceedings of the 3rd workshop on games and NLP
14. Misra A, Anand P, Tree JEF, Walker MA (2015) Using summarization to discover argument facets in online idealogical dialog. In: NAACL HLT 2015, The 2015 conference of the north

american chapter of the association for computational linguistics: human language technologies, Denver, Colorado, USA, May 31 - June 5, 2015. http://aclweb.org/anthology/N/N15/N15-1046.pdf
15. Misra A, Ecker B, Walker MA (2016) Measuring the similarity of sentential arguments in dialogue. In: Proceedings of the SIGDIAL 2016 conference, the 17th annual meeting of the special interest group on discourse and dialogue, 13–15 September 2016, Los Angeles, CA, USA. http://aclweb.org/anthology/W/W16/W16-3636.pdf
16. Oraby S, Harrison V, Reed L, Hernandez E, Riloff E, Walker MA (2016) Creating and characterizing a diverse corpus of sarcasm in dialogue. In: Proceedings of the SIGDIAL 2016 conference, The 17th annual meeting of the special interest group on discourse and dialogue, 13–15 September 2016, Los Angeles, CA, USA, pp 31–41. http://aclweb.org/anthology/W/W16/W16-3604.pdf
17. Oraby S, Reed L, Compton R, Riloff E, Walker MA, Whittaker S (2015) And that's a fact: distinguishing factual and emotional argumentation in online dialogue. In: Proceedings of the 2nd workshop on argumentation mining, ArgMining@HLT-NAACL 2015, June 4, 2015, Denver, Colorado, USA, pp 116–126. http://aclweb.org/anthology/W/W15/W15-0515.pdf
18. Price P, Hirschman L, Shriberg E, Wade E (1992) Subject-based evaluation measures for interactive spoken language systems. In: Proceedings of the workshop on speech and natural language, pp 34–39. Association for Computational Linguistics
19. Ritter A, Cherry C, Dolan WB (2011) Data-driven response generation in social media. In: Proceedings of the conference on empirical methods in natural language processing. Association for Computational Linguistics
20. Serban IV, Lowe R, Charlin L, Pineau J (2015) A survey of available corpora for building data-driven dialogue systems. CoRR **abs/1512.05742**
21. Sordoni A, Galley M, Auli M, Brockett C, Ji Y, Mitchell M, Nie JY, Gao J, Dolan B (2015) A neural network approach to context-sensitive generation of conversational responses. arXiv preprint http://arxiv.org/abs/1506.06714
22. Swanson R, Gordon AS (2008) Say anything: a massively collaborative open domain story writing companion. In: Interactive storytelling. Springer, pp 32–40
23. Tsiakoulis P, Gasic M, Henderson M, Plannels J, Prombonas J, Thomson B, Yu K, Young S, Tzirkel E (2012) Statistical methods for building robust spoken dialogue systems in an automobile. In: in 4th international conference on applied human factors and ergonomics
24. Vinyals O, Le Q (2015) A neural conversational model. arXiv preprint http://arxiv.org/abs/1506.05869 (2015)
25. Walker MA, Litman DJ, Kamm CA, Abella A (1997) Paradise: a framework for evaluating spoken dialogue agents. In: Proceedings of the eighth conference on european chapter of the association for computational linguistics, pp 271–280. Association for Computational Linguistics
26. Walker MA, Passonneau R, Boland JE (2001) Quantitative and qualitative evaluation of darpa communicator spoken dialogue systems. In: Proceedings of the 39th annual meeting on association for computational linguistics, pp 515–522. Association for Computational Linguistics

ChaD: Chat-Oriented Dialog Systems

Alexander I. Rudnicky

Abstract Historically, conversational systems have focused on goal-directed interaction and this focus defined much of the work in the field of spoken dialog systems. More recently researchers have started to focus on non- goal-oriented dialog systems often referred to as "chat" systems. We refer to these as Chat-oriented Dialog (ChaD) systems. ChaD systems are not task-oriented and focus on what could be described as social conversation where the goal is to interact with a human interlocutor while maintaining an appropriate level of engagement. Research to date has identified a number of techniques that can be used to implement working ChaDs but it has also highlighted important limitations. This note describes key ChaD characteristics and proposes a research agenda.

Keywords Conversational systems · Chatbots · Evaluation

1 Introduction

Historically, conversational systems have focused on goal-directed interaction (e.g. [3]) and this focus defined much of the work in the field of spoken dialog systems. More recently researchers have started to focus on non- goal-oriented dialog systems [2, 9, 11, 13] often referred to as "chat" systems. We refer to these as Chat-oriented Dialog (ChaD) systems. ChaDs differ from "chatbots", systems that have recently gained significant attention. The latter have evolved from on-line, text-based, chat systems, originally supporting human-human conversations and more recently finding acceptance as a supplement to call centers. Chatbots have begun to resemble goal-oriented dialog systems in their use of natural language understanding and generation, as well as dialog management. Speech-based implementations have begun to appear in commercial applications (for example, see the recent Conversational Interaction Conference [1]).

A. I. Rudnicky (✉)
Carnegie Mellon University, Pittsburgh, PA 15213, USA
e-mail: air@cs.cmu.edu

M. Eskenazi et al. (eds.), *Advanced Social Interaction with Agents*, Lecture Notes in Electrical Engineering 510, https://doi.org/10.1007/978-3-319-92108-2_7

By contrast, CHAD systems are not meant to be task-oriented and instead focus on what can be described as social conversation where the goal is to interact while maintaining an appropriate level of engagement with a human interlocutor. One consequence is that many of the metrics that have been developed for task-oriented systems no longer are appropriate. For example, the purpose of a task dialog is to achieve a satisfactory goal (e.g. provide some unit of information) and to do so as rapidly and accurately as practical. The goal in CHAD on the other hand is almost the opposite: keep the human engaged in a conversation for an extended period of time, with no concrete goal in mind other than producing a pleasant experience or developing a social relationship. In this note we will use the term CHAD to refer to this latter class of system.

2 Contemporary CHAD systems

CHAD systems have been of interest for many years. An early, and influential, example was ELIZA [12], more recent examples include A.L.I.C.E. [10] and Tank the Roboceptionist [6, 8]. These systems are rule-based; that is, responses are generated based on text patterns hand-crafted to match user inputs, with limited use of context and history. More recent attempts have take a corpus-based approach [2, 13], where some source of authored conversations (say movie dialogues) is indexed and features of user inputs are used as retrieval keys. Deep Learning approaches, e.g. [9, 11, 13] have been successfully implemented for this task. A persistent limitation of these approaches is that they tend to reduce to an equivalent of question answering: the immediately preceding user input and perhaps the previous system turn are part of the retrieval key. This narrow window necessarily reduces continuity and users struggle to follow a conversational thread.

A key research question is how to create conversations that are initiated and develop in ways similar to those found in human-human conversation. Human (social) conversations exhibit structure that guides their evolution over time. This structure includes elements familiar to participating interlocutors such as conventions for engagement and disengagement, a succession of topics (with heuristics for identifying promising successors), as well as monitoring engagement level and picking strategies for managing topic transitions. Good conversational management also implies sophisticated use of memory, both within the conversation to support back references and general world knowledge (or at the very least domain-specific knowledge), to generate proper turn succession. The implementation of such capabilities is not well understood at this time.

3 Research Challenges

Work to date has identified a number of techniques that can be used to implement working CHADs but it has also highlighted some important limitations. These include the following:

- **Continuity**: Response generation techniques have focused primarily on database retrieval; even deep-learning approaches have that character. In this approach a response is generated based on a very short history, typically the previous human input. The unfortunate consequence is that system behavior begins to resemble questions-answering, for example exhibiting sudden topic changes when appropriate responses cannot be retrieved. Missing is some sense of continuity and history that would allow the system to build momentum for a topic, recognize topic re-introductions, decide when and how to change the topic or otherwise communicate that it is paying attention to the conversation. At the same time it places a burden on the human participant: their contribution becomes a succession of questions.
- **Dialogue Strategies:** Humans are quite adept at managing their conversational behavior, by monitoring interlocutor engagement [14] and by maintaining higher-level goals [5, 7]. Current CHAD approaches are still at an early stage and do not yet exhibit the natural flow one is accustomed to observe in human conversations. A related challenge is how to effectively interleave social and task goals in a single conversation and be able to coordinate strategies across both streams to create a fluent performance that is consistent with human expectations.
- **Evaluation**: Current evaluation techniques for CHAD rely on human judgment. This indicates that we do not as yet understand how to quantify conversational success. This is a serious problem on several levels. The lack of a clear, automatically computable metric hinders development of corpus-based approaches (the dominant scheme in current research) which lack suitable objective (i.e. loss) functions to guide learning. But the problem extends to human judgment, as currently proposed rating schemes exhibit wide cross-judge variation [4], making it difficult to understand underlying phenomena.

Irrespective of these challenges CHAD continues to attract a significant amount of interest and is generating an increasing body of research. The problem of creating fluent conversation provides the opportunity to address a broader range of phenomena than ones that have been the focus in task-oriented systems and develop a richer understanding of natural human communication.

References

1. AVIOS: Conversational interaction conference. San Jose, CA (2017) http://www.conversationalinteraction.com/
2. Banchs RE, Li H (2012) IRIS: a chat-oriented dialogue system based on the vector space model. In: Proceedings of the ACL 2012 system demonstrations, pp 37–42. Association for Computational Linguistics
3. Bohus D, Rudnicky AI (2009) The RavenClaw dialog management framework: architecture and systems. Comput Speech Lang 23(3):332–361
4. Charras F, Dubuisson Duplessis G, Letard V, Ligozat AL, Rosset S (2016) Comparing system-response retrieval models for open-domain and casual conversational agent. In: Second workshop on Chatbots and conversational agent technologies
5. Foster ME, Gaschler A, Giuliani M, Isard A, Pateraki M, Petrick RP (2012) Two people walk into a bar: Dynamic multi-party social interaction with a robot agent. In: Proceedings of the 14th ACM international conference on multimodal interaction, ICMI '12. ACM, New York, NY, USA, pp 3–10. https://doi.org/10.1145/2388676.2388680
6. Gockley R, Bruce A, Forlizzi J, Michalowski M, Mundell A, Rosenthal S, Sellner B, Simmons R, Snipes K, Schultz AC et al (2005) Designing robots for long-term social interaction. In: 2005 IEEE/RSJ international conference on intelligent robots and systems, 2005. (IROS 2005). IEEE, pp 1338–1343
7. Hiraoka T, Georgila K, Nouri E, Traum DR, Nakamura S (2015) Reinforcement learning in multi-party trading dialog. In: SIGDIAL conference
8. Makatchev M, Fanaswala I, Abdulsalam A, Browning B, Ghazzawi W, Sakr MF, Simmons RG (2010) Dialogue patterns of an arabic robot receptionist. In: HRI
9. Ritter A, Cherry C, Dolan WB (2011) Data-driven response generation in social media. In: Proceedings of the conference on empirical methods in natural language processing, pp 583–593. Association for Computational Linguistics
10. Shawar BA, Atwell E (2003) Using dialogue corpora to train a chatbot. In: Proceedings of the Corpus Linguistics 2003 conference, pp 681–690
11. Vinyals O, Le QV (2015) A neural conversational model. In: ICML deep learning workshop. http://arxiv.org/pdf/1506.05869v3.pdf
12. Weizenbaum J (1966) ELIZA—a computer program for the study of natural language communication between man and machine. Commun. ACM 9(1): 36–45. https://doi.org/10.1145/365153.365168
13. Yu Z, Papangelis A, Rudnicky A (2015) TickTock: a non-goal-oriented multimodal dialog system with engagement awareness. In: Proceedings of the AAAI spring symposium
14. Yu Z, Xu Z, Black AW, Rudnicky AI (2016) Strategy and policy learning for non-task-oriented conversational systems. In: 17th annual meeting of the special interest group on discourse and dialogue, vol 2, p 7

Part II
Multi-domain Dialogue Systems

Domain Complexity and Policy Learning in Task-Oriented Dialogue Systems

Alexandros Papangelis, Stefan Ultes and Yannis Stylianou

Abstract In the present paper, we conduct a comparative evaluation of a multitude of information-seeking domains, using two well-known but fundamentally different algorithms for policy learning: GP-SARSA and DQN. Our goal is to gain an understanding of how the nature of such domains influences performance. Our results indicate several main domain characteristics that play an important role in policy learning performance in terms of task success rates.

1 Introduction

As we move towards intelligent dialogue-based agents with increasingly broader capabilities, it becomes imperative for these agents to be able to converse over multiple topics. A few approaches have been proposed in the recent literature [1–3], however it is not clear which approach will scale to real-world applications that involve multiple large domains. In this work, aiming to design better performing and more scalable dialogue policy learning algorithms, we look into which domain factors result in high domain complexity with respect to learned dialogue policy performance. In particular, we look closely into information-seeking domains, and to our knowledge this is the first attempt to systematically examine the relation between domain factors and the achieved dialogue performance.

To prevent the emergence of factors that can be attributed to the training algorithm or condition (single- or multi-domain), we select two fundamentally different algorithms proven to be robust in each condition, namely GP-SARSA [4] and DQN [5],

A. Papangelis (✉) · Y. Stylianou
Toshiba Research Europe, Cambridge, UK
e-mail: alex.papangelis@crl.toshiba.co.uk

Y. Stylianou
e-mail: yannis.stylianou@crl.toshiba.co.uk

S. Ultes
Department of Engineering, University of Cambridge,
Cambridge, UK
e-mail: su259@cam.ac.uk

M. Eskenazi et al. (eds.), *Advanced Social Interaction with Agents*, Lecture Notes in Electrical Engineering 510, https://doi.org/10.1007/978-3-319-92108-2_8

Table 1 Characteristics of the investigated domains. Slots, avg. values, coverage and entropy only reflect the system requestable slots. DB size refers to the number of DB records and Avg. Values is the average number of values per slot

Domain	Sys. Req. slots	Avg. values	DB size	Coverage	Sum entropy
CR	3	10.66	110	0.1888	4.8188
CH	5	3.2	33	0.1	4.4606
CS	2	6	21	0.3428	2.9739
L11	11	4.55	123	0.0002	12.3388
L6	6	3	123	0.0556	6.1471
TV	6	4.5	94	0.0187	5.5966
SH	6	9.5	182	0.0028	7.0026
SR	6	23.5	271	0.0002	10.9190

and investigate whether their performance is influenced by specific domain characteristics, i.e., the domain complexity.

Finding a measure for the complexity of information-seeking scenarios has previously been the focus of research. While most of the relevant work focuses on language-related effects [6–10], e.g., syntactic or terminological complexity, we are focusing on the structural or semantic complexity of the application domain. This has previously been studied by Pollard and Biermann [11] who defined a schema for calculating a complexity measure based on entropy. Building upon that and using additional features, we aim at providing experimental evidence for this by identifying the domain factors which contribute the most to the resulting success rate of the dialogue policy employed.

The remainder of the paper is organised as follows: the notion of information seeking dialogue and the investigated domains are presented in Sect. 2, followed by the algorithms used for policy learning in Sect. 3. In Sect. 4, the results are presented mapping domain characteristics to task success rates before concluding in Sect. 5 including an outlook on future work.

2 Information-Seeking Domains

We formally define information-seeking (or slot-filling) domains (ISD) for dialogue as tuples $\{S, V, A, D\}$, where $S = \{s_0, ..., s_N\}$ is a set of slots, V is a set of values that each slot can take $s_i \in V_i$, A is a set of system dialogue acts of the form $intent(s_0 = v_0, ..., s_k = v_k)$, and D represents a database of items, whose characteristics can be described by the slots. Slots may be further categorised as *system requestable* i.e. slots whose value the system may request, *user requestable* i.e. slots whose value the user may request, and *informable* i.e. slots whose value the user may provide.

For our evaluation, we have selected a number of domains of different complexity, namely: Cambridge Restaurants (CR), Cambridge Hotels (CH), Cambridge Shops

(CS), Laptops 11 (L11), Laptops 6 (L6), Toshiba TVs (TV), San Francisco Hotels (SH), and San Francisco Restaurants (SR), where L11 is an extended version of L6. Table 1 lists relevant characteristics of the domains we use in our evaluation. *Coverage* refers to the ratio of the unique set of available combinations of slot-value pairs of the items in the database (D) over all possible combinations, taking into account *system requestable* slots only as they are used for constraining the search:

$$Coverage(D) = \frac{|unique(D)|}{\prod_{s}^{|S_{sr}|} V_s} \quad (1)$$

where $unique(D)$ is the set of unique items in the database with respect to system requestable slots, S_{sr} is the set of system requestable slots and V_s is the set of available values of each $s \in S_{sr}$. In addition, we computed each domain's slot entropies and normalised slot entropies:

$$H(S) = -\sum_{i=1}^{|V_i|} p(S = s_i) \, \log p(S = s_i) \quad (2)$$

Since the number of slots and number of values per slot varies across domains, we computed descriptive statistics (min, max, average, st.dev., and sum) for each of these features.

3 Dialogue Policy Learning

Statistical Dialogue Management. Using statistical methods for dialogue policy learning (and consequently dialogue management) has prevailed in the state of the art for many years. Partially Observable Markov Decision Processes (POMDP) have been preferred in dialogue management due to their ability to handle uncertainty, which is inherent in human communication. Concretely, a POMDP is defined as a tuple $\{S, A, T, O, \Omega, R, \gamma\}$, where S is the state space, A is the action space, $T: S \times A \rightarrow S$ is the transition function, $O: S \times A \rightarrow \Omega$ is the observation function, Ω is a set of observations, $R: S \times A \rightarrow \Re$ is the reward function and $\gamma \in [0, 1]$ is a discount factor of the expected cumulative rewards $J = E[\sum_t \gamma^t R(s_t, a_t)]$. A policy $\pi: S \rightarrow A$ dictates which action to take from each state. An optimal policy $\pi^\star$ selects an action that maximises the expected reward of the POMDP, J. Learning in RL consists exactly of finding such optimal policies; however, due to state-action space dimensionality, approximation methods are needed for practical applications.

GP-SARSA (GPS) [4] is an online RL algorithm that uses Gaussian processes to approximate the Q function. It has been successfully used to learn dialogue policies [12, e.g.] and therefore was a strong candidate for our evaluation.

Deep Q-Networks (DQN) [5] is an RL algorithm that uses deep neural networks (DNN) to approximate the Q function. In this work, we apply DQN on a multi-domain dialogue manager using domain-independent input features [3]. A simple

Table 2 Average success rates for each domain and learning algorithm

Algorithm	CR	CH	CS	L11	L6	TV	SH	SR
GPS	86.9	69.2	91	50.5	63.8	79.8	65.6	57.7
DQN	81.9	69.5	85.9	74.8	68.9	84.3	76.8	71.7

fully connected feed-forward network is used with two layers of 60 and 40 nodes and sigmoid activations.

We trained the above algorithms with the PyDial toolkit [13], using the following reward functions: for the GPS, we assign a turn penalty of -1 for each turn and a reward of +20 at the end of each successful dialogue. For the DQN, we use the same turn penalty, but a -200 penalty for unsuccessful dialogues and a +200 reward for successful dialogues, divided by the number of active domains seen during the training dialogue. We used higher rewards and penalties in this case to account for the longer dialogues when having more than one domain. We construct the summary actions as follows: $request(slot_i)$, $confirm(slot_i)$, and $select(slot_i)$ for all system requestable slots, plus the following actions without slot arguments: *inform, inform_byname, inform_alternatives, inform_requested, bye, repeat, request_more, restart*. The action space of each domain therefore is $|A| = 3|S_{sr}| + 8$. Arguments for the summary actions are instantiated in the mapping from summary to full action space.

4 Evaluation and Analysis

To see the effects of the various domain characteristics on performance, we trained GPS policies on each domain, recording the dialogue success rates averaged over 10 runs of 1,000 training dialogue/100 evaluation dialogue cycles. Training and evaluation was conducted in simulation using an updated version of the simulated user proposed in [14] with a semantic error rate of 15% (probability by which the user's act is distorted in terms of slots and/or values). In order to see if such effects may indeed be attributed to the domain and not to the algorithm, one domain-independent DQN policy was trained with dialogues from all of the available domains using a domain-independent dialogue state representation. By having 2–4 active domains in each dialogue (randomly sampled), the DQN effectively was trained on more data than each GPS was. The evaluation of the DQN, however, was done on single domains. Again the dialogue success rates were averaged over 10 runs of 1,000 training dialogue/100 evaluation dialogue cycles.

Table 2 shows the average dialogue success rates for the two algorithms we evaluated. The task success rates of both algorithms clearly show similar performance on the respective domains; in fact, both results are highly correlated ($\rho = 0.8$). Although not in the focus of this work, it may also be seen that the DQN policy is able to learn

Table 3 The top-3 standardised coefficients ordered by increasing p value of the linear regression for the statistically significant stepwise linear regression models. Bold represents $p < 0.01$

Mdl1	Mdl2	Mdl3	Mdl4	Mdl5	Mdl6	Mdl7	Mdl8
SumEnt	**SysReq**	**Coverage**	**SumNEnt**	**DBItems**	**StdVal**	**AvgVal**	**StdEnt**
−0.867	−1.879	0.805	−0.948	−0.544	−0.574	−0.537	−0.547
SysReq	UsrReq	MinEnt	**MaxVal**	**StdVal**	AvgEnt	AvgEnt	MinEnt
−0.468	1.084	−0.732	−0.595	−0.574	0.381	0.571	−0.421
Coverage	MaxNEnt	SumNEnt	StdVal	**StdEnt**	MaxNEnt	MaxNEnt	MaxEnt
0.382	−0.194	−0.457	1.440	−0.547	0.268	0.313	−0.189

solutions that are more general, thus mitigating the effects of hard-to-train domains (e.g. CH, SR or L11) to some degree. Still, the GPS policy performs better in other domains (e.g. CR and CS). All in all, this shows that although both algorithms have fundamentally different characteristics, the resulting success rates are highly correlated.

In order to identify domain characteristics which correlate with the performance of a learned policy, we analysed the results by running stepwise linear regressions to investigate which of the domain characteristics (independent variables) better explained the success rate (dependent variable). Table 3 shows the coefficients of the GPS models. The most influential one seems to be ***sum of slot entropies*** (Model 1—Mdl1), which explains 75.1% of the variance ($p < 0.005$). If we remove this characteristic and run the regression again (Mdl2), the number of ***system requestable slots*** explains 75% of the variance ($p < 0.005$). In a further step of linear regression having removed the latter characteristic, ***coverage*** (Mdl3) appears to explain 64.8% of the variance ($p < 0.01$). The rest of the models yield the variables shown in Table 3, each of which explains about 80% of the variance in the respective model, with $p < 0.01$. All of the above are drawn from the GPS results and calculated at a 95% significance level.

After observing our data given the above analysis, it seems that as the *sum of entropies* increases, the algorithm's performance drops as there are more values per system requestable slot to explore. Regarding the second most influential factor, as the number of *system requestable slots* increases, the dialogue success rate (given 1,000 training dialogues) decreases, as was expected. This is because the size of the model representing the policy is directly related to the number of *system requestable slots*, and as the latter increase we need a larger model to represent the policy and therefore more training dialogues. The inverse trend holds for the third most influential factor, *coverage*; as it increases, so does the dialogue success rate. The reason for this may be the fact that with large *coverage* it is easier for the policy to learn actions with high discriminative power with respect to DB search, when compared to the case of small *coverage*.

For the DQN, the learning algorithm has access to training examples from a variety of domains (since we train a single domain-independent policy model), and this results in effects of domains with higher coverage, entropy or many system requestable slots being averaged out in terms of performance. However, we still observe trends similar to GPS in terms of dialogue success across the domains and indeed, as mentioned above, the success rates of the two conditions are highly correlated ($\rho = 0.8$). This will be investigated further in future work, by evaluating more domains.

5 Conclusion

We presented an analysis of the characteristics of ISD that have an impact on the performance of dialogue policy learning algorithms using two different algorithms. Our results show that the sum of each system-requestable slot's entropy plays a significant role, along with the number of system requestable slots, database coverage, and other characteristics. These results will help judge the difficulty of finding a well-performing dialogue policy as well as the design of policy learning algorithms.

Of course, our analysis depends on how we define the dialogue as an optimisation problem. As future work, we plan to evaluate more learning algorithms on a larger number of domains (primarily information-seeking), aiming at designing an abstract domain generator that will create various classes of benchmark ISDs to be used when evaluating new policy learning algorithms.

References

1. Gašić M, Mrkšić N, Rojas-Barahona LM, Su P-H, Ultes S, Vandyke D, Wen T-H, Young S (2016) Dialogue manager domain adaptation using gaussian process reinforcement learning. Comput Speech Lang
2. Cuayáhuitl H, Yu S, Williamson A, Carse J (2016) Deep reinforcement learning for multi-domain dialogue systems. arXiv preprint arXiv:1611.08675
3. Papangelis A, Stylianou, Y (2016) Multi-domain spoken dialogue systems using domain-independent parameterisation. In: Domain adaptation for dialogue agents
4. Engel Y, Mannor S, Meir R (2005) Reinforcement learning with gaussian processes. In: Proceedings of the 22nd ICML. ACM, pp 201–208
5. Mnih V, Kavukcuoglu K, Silver D, Rusu AA, Veness J, Bellemare MG, Graves A, Riedmiller M, Fidjeland AK, Ostrovski G et al (2015) Human-level control through deep reinforcement learning. Nature 518(7540):529–533
6. Remus R (2012) Domain adaptation using domain similarity- and domain complexity-based instance selection for cross-domain sentiment analysis. In: 2012 IEEE 12th international conference on data mining workshops, Dec 2012, pp 717–723
7. Freitas A, Sales JE, Handschuh S, Curry E (2015) How hard is this query? Measuring the semantic complexity of schema-agnostic queries. In: IWCS 2015, p 294
8. Grubinger M, Leung C, Clough P (2005) Linguistic estimation of topic difficulty in cross-language image retrieval. In: Workshop of the cross-language evaluation forum for European languages. Springer, pp 558–566
9. Sebastiani F (1994) A probabilistic terminological logic for modelling information retrieval. In: SIGIR94. Springer, pp 122–130

10. Bagga A, Biermann AW (1997) Analyzing the complexity of a domain with respect to an information extraction task. In: Proceedings of the tenth international conference on research on computational linguistics (ROCLING X), pp 175–194
11. Pollard S, Biermann AW (2000) A measure of semantic complexity for natural language systems. In: Proceedings of the NAACL SSCNLPS, Stroudsburg, PA, USA, pp 42–46
12. Gašić M, Breslin C, Henderson M, Kim D, Szummer M, Thomson B, Tsiakoulis P, Young S (2013) On-line policy optimisation of bayesian spoken dialogue systems via human interaction. In: 2013 IEEE international conference on acoustics, speech and signal processing (ICASSP). IEEE, pp 8367–8371
13. Ultes S, Rojas-Barahona L, Su PH, Vandyke D, Kim D, Casanueva I, Budzianowski P, Mrkšić N, Wen TH, Gašić M, Young S (2017) Pydial: a multi-domain statistical dialogue system toolkit. In: ACL 2017 Demo, Vancouver. ACL
14. Schatzmann J, Young SJ (2009) The hidden agenda user simulation model. IEEE Trans Audio Speech Lang Process 17(4):733–747

Single-Model Multi-domain Dialogue Management with Deep Learning

Alexandros Papangelis and Yannis Stylianou

Abstract We present a Deep Learning approach to dialogue management for multiple domains. Instead of training multiple models (e.g. one for each domain), we train a single domain-independent policy network that is applicable to virtually any information-seeking domain. We use the Deep Q-Network algorithm to train our dialogue policy, and evaluate against simulated and paid human users. The results show that our algorithm outperforms previous approaches while being more practical and scalable.

Keywords Spoken dialogue system · Multi domain · Deep learning
Dialogue policy learning

1 Introduction

With the proliferation of intelligent agents, assisting us on various daily tasks (customer support, information seeking, hotel booking, etc.), several challenges of Spoken Dialogue Systems (SDS) become increasingly important. Following the successful application of statistical methods for SDS—casting the dialogue problem as a Partially Observable Markov Decision Process (POMDP) and applying reinforcement learning (RL) algorithms—several approaches have been proposed to tackle these challenges as well as to reduce the development effort. One such challenge is the ability of an SDS to converse on multiple topics or domains.

Statistical Dialogue Management. POMDPs have been preferred in dialogue management due to their ability to handle uncertainty, which is inherent in human

---

A. Papangelis (✉) · Y. Stylianou
Toshiba Research Europe, Cambridge, UK
e-mail: alex.papangelis@crl.toshiba.co.uk

Y. Stylianou
e-mail: yannis.stylianou@crl.toshiba.co.uk

Y. Stylianou
Department of Computer Science, University of Crete, Heraklion, Greece

M. Eskenazi et al. (eds.), *Advanced Social Interaction with Agents*, Lecture Notes in Electrical Engineering 510, https://doi.org/10.1007/978-3-319-92108-2_9

communication. A POMDP Dialogue Manager (DM) typically receives an n-best list of language understanding hypotheses, which are used to update the belief state (reflecting an estimate of the user's goals). Using RL, the system selects a response that maximises the long-term return of the system. This response is typically selected from an abstract action space and is converted to text through language generation. Concretely, a POMDP is defined as a tuple $\{S, A, T, O, \Omega, R, \gamma\}$, where S is the state space, A is the action space, $T : S \times A \rightarrow S$ is the transition function, $O : S \times A \rightarrow \Omega$ is the observation function, Ω is a set of observations, $R : S \times A \rightarrow \Re$ is the reward function and $\gamma \in [0, 1]$ is a discount factor of the expected cumulative rewards $J = E[\sum_t \gamma^t R(s_t, a_t)]$. A policy $\pi : S \rightarrow A$ dictates which action to take at each state. An optimal policy $\pi^\star$ selects an action that maximises the expected returns of the POMDP, J. Learning in RL consists of finding such optimal policies; however, due to state-action space dimensionality, approximation methods need to be used for practical applications. One such method is to perform learning on summary spaces [18, e.g.].

Relevant Work. Recently, Gašić et al. [4, 5] proposed the use of a hierarchical structure to train generic dialogue policies that can then be refined when in-domain data become available. A Bayesian Committee Machine (BCM) over multiple dialogue policies (each trained on one domain) decides which policy can better handle the user's utterance, and delegates control to that policy. Using this structure, the system can deal with multi-domain dialogues. In order to adapt to new speakers with dysarthria, Casanueva et al. [1] explore ways of transferring data from known speakers, to improve the cold-start performance of a SDS. In particular, when a new speaker interacts with the system, they propose a way to select data from speakers that are similar to the new speaker, and weigh them appropriately.

In [3], the authors use Deep RL to train a policy network which takes as input noisy text (thus bypassing Spoken Language Understanding) and outputs the system's action. The latter can either be simple (e.g. inform(·)) or composite (e.g. a sub-dialogue ≡ domain). The internal structure of the model is a network of Deep-Q policy networks, each of which is learning a dialogue policy for a given domain, plus one network for general dialogues (e.g. greetings etc.). The authors train Naive Bayes classifiers to identify valid actions from each dialogue state and they train a Support Vector Machine (SVM) to select the appropriate domain. They evaluate their system on a two-domain information seeking task, for hotels and restaurants. Our work is different in that we train a single Deep-Q network that is able to operate across domains, therefore making it much more scalable, since in Cuayahuitl et al. [3] work it is necessary to add a network for each domain, and the actions in the output are domain-specific.

Other scholars tackle the problem of adapting to new or known users over time, or focus on different parts of the dialogue system (e.g. [2, 7, 8, 14, 16, 20]). Our approach, however, does not rely on complicated transfer learning methods [12, 13] but instead on modelling the generic class of information seeking dialogues by abstracting away from the specifics of each domain. In prior work, Wang et al. [17] proposed a domain-independent summary space (applicable to information-seeking dialogues) onto which a learning algorithm can operate. This allows policies trained on one domain to be transferred to other, unseen domains. In [10], we proposed to apply Wang et al. domain transfer method to design a multi-domain dialogue manager. Here we extend this work by applying Deep Q-Networks and show that this outperforms the previous, GP-SARSA-based multi-domain SDS.

2 Multi-Domain Dialogue Management

Domain Independent Parameterisation (DIP) [17] is a method that maps the (belief) state space into a feature space of size N, that is independent of the particular domain: $\Phi_{DIP}(s, l, a) : S \times L \times A \rightarrow \Re^N, s \in S, l \in L, a \in A$, where L is the set of slots (including a 'null' slot for actions such as *hello*). Φ_{DIP} therefore extracts features for each slot, given the current belief state, and depends on A in order to allow for different parameterisations for different actions. This allows us to define a fixed-size domain-independent space, and policies learned on this space can be used in various domains, in the context of information-seeking dialogues. As shown in [10], we can take advantage of DIP to design efficient multi-domain dialogue managers, the main benefit being that we learn a single, domain-independent policy model that can be applied to information-seeking dialogues. Aiming to further improve the efficiency and scalability of such dialogue managers, we propose to use a variant of DQN [9] to optimise the multi-domain policy. To this end, we use a two-layer feed-forward network (FFN) to approximate the Q function.

Achieving Domain Independence. To be completely domain-independent, we need to define a generic action space $\mathcal{A}$ for information seeking problems. For our experiments, we include the following system actions: *hello, bye, inform, confirm, select, request, request_more, repeat*. The policy thus operates on the $\Phi_{DIP} \times \mathcal{A}$ space, instead of the original belief-action space. By operating in this parameter space and letting the policy decide which action to take next as well as which slot the action refers to, we achieve independence in terms of both slots and actions, as long as the actions of any domain can be represented as functions of $\mathcal{A} \times L$, where L are the domain's slots including the 'null' slot. The learned policy therefore decides which action to take by maximising over both the action and the slots:

$$a_{t+1} = argmax_{l,a}\{Q[\Phi_{DIP}(s_t, l, a), a]\} \quad (1)$$

where $a_{t+1} \in \mathcal{A} \times L$ is the selected summary action, s_t is the belief state at time t, and $a \in \mathcal{A}$. To approximate the Q function, we use a 2-layered FFN with 60 and 40 hidden nodes, respectively. The input layer receives the DIP feature vector $\Phi_{DIP}(s, l, a)$ and the output layer is of size $\mathcal{A}$; each output dimension can be interpreted as $Q[\Phi_{DIP}(s, l, a), a]$:

$$\overrightarrow{Q}(\Phi_{DIP}(s_t, l, a)) \approx softmax(W_k^M x_k^{M-1} + b^M) \quad (2)$$

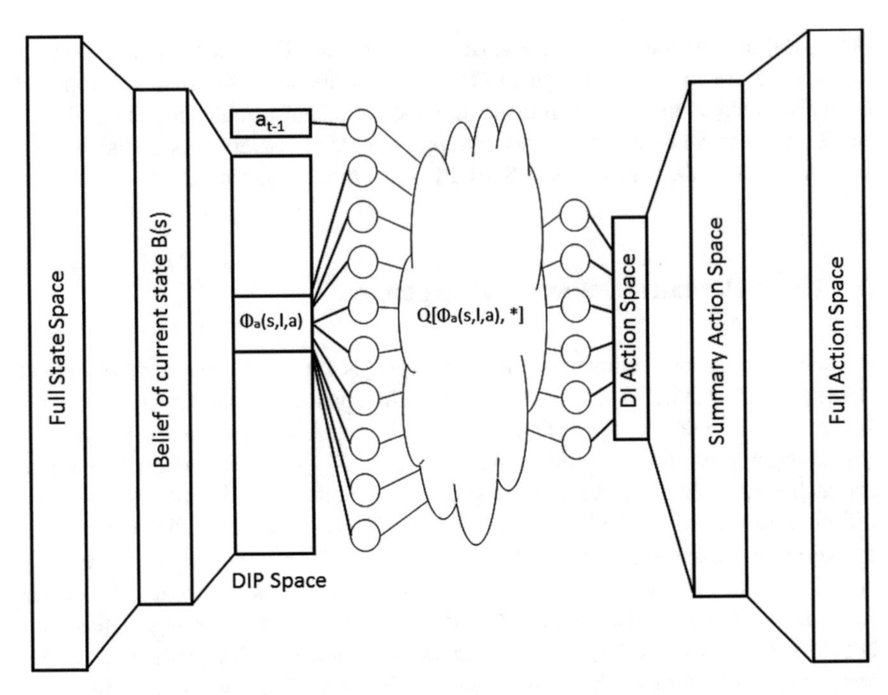

Fig. 1 The architecture of our DNN-based multi-domain dialogue manager. It should be noted that any policy learning algorithm can be used in place of DQN

where $\overrightarrow{Q}(\Phi_{DIP}(s_t, l, a))$ is a vector of size $|\mathcal{A}|$, W^m are the weights of the mth layer (out of M layers in total) for nodes k, x_k^m holds the activations of the mth layer, where $x^0 = \Phi_{DIP}(s_t, l, a)$, and b^m are the biases of the mth layer. To generate the summary system action, we simply combine the selected slot and action from equation (1).

The architecture of the system is shown in Fig. 1. We train the model with DQN with experience replay [9], standardizing the input vector of each minibatch to ensure that it follows the same distribution over time. However, we introduce a small bias in the minibatch sampling, towards datapoints with infrequent rewards. In particular, we sample datapoints d from an exponential-like distribution with a small λ value, taking into account the probability of a datapoint to occur in the experience pool, given a reward r, similarly to [11]. If done efficiently, this only introduces a linear factor in the algorithm's time complexity while considerably improving performance and robustness. We used a pool of 1,000 datapoints and a minibatch of 100.

3 Evaluation

Using the PyDial system [15], we trained the DQN-based DM in simulation on four domains: Cambridge Attractions (CA), Cambridge Shops (CS), Cambridge Hotels (CH), and Cambridge Restaurants (CR) using a topic tracker to switch between them, for SLU and NLG purposes. We allowed 1,000 training dialogues (training for one epoch after every ten dialogues) and 1,000 evaluation dialogues while varying the semantic error rate. To assess performance, we compare against a DIP-based DM trained with GPS and a baseline of policies trained with GPS in a single-domain setup (GPS-IND). For the DIP-based algorithms we allow 10 dialogue turns per active domain; this means that the multi-domain dialogues are longer than the single-domain ones, as the active domains at each dialogue and can range from 2 to 4. However, in the multi-domain condition, each domain is active on average in less than 750 dialogues. We therefore train the single-domain policies allowing 40 turns per dialogue, for 750 dialogues.

We then evaluate the multi-domain DMs against a BCM-based DM trained under the same conditions, by conducting a small human user trial, as due to certain restrictions we were not able to conduct crowd-sourced experiments.

Results. Table 1 shows the results of the experiments in simulation, where we varied the semantic error rate, averaged over 10 training/testing runs. We can see that the DQN-based DM outperforms GPS-DIP on multiple domains, as it is more robust to higher error rates (e.g. at the CH domain). Both DIP DMs outperform the baseline in the no-noise condition as they are able to learn more general policies and mitigate effects of harder-to-train domains (e.g. CH). In the presence of noise, GPS-DIP does not seem to cope very well, contrary to DQN-DIP which seems to fare much better in deteriorating conditions even though both algorithms use the same input.

We conducted a small user trial (30 interactions) comparing the DIP DMs and a BCM DM trained under the same conditions for 1,000 multi-domain dialogues. We asked participants to engage with the SDS in multi-domain dialogues (2–4 domains simultaneously active). Success was computed by comparing the retrieved item with the participant's goals for that session. In the end, participants were asked how they would rate the dialogue overall, and had to provide an answer from 1 (very bad) to 5 (excellent). Table 2 shows the results, where we can see that DQN performs very closely to BCM. Even though we can't draw strong conclusions from 30 interactions, it is evident that DQN using a single policy model performs at least as well as BCM, which uses one policy model for each domain. More trials will be conducted in the near future, including the DM proposed in [3].

Table 1 Dialogue Success rates for the three dialogue managers under evaluation

	DQN-DIP				GPS-DIP				GPS-IND			
Error	0%	15%	30%	45%	0%	15%	30%	45%	0%	15%	30%	45%
CA	95.65	94.2	88.41	79.17	95.5	94.44	77.78	28.17	87.1	78.9	68.7	59.2
CS	95.45	92.96	90.48	76.47	92.59	92	84.62	54.55	89.6	87.2	80.4	71.9
CH	81.25	71.62	64.47	45	88.89	26.92	16.67	10.61	64.6	47.9	35	21.8
CR	92.86	90.62	87.88	71.23	82.61	77.78	61.9	36.11	86.5	78.4	74	59.2
AVG	91.30	87.35	82.81	67.97	89.90	72.79	60.24	32.36	81.95	73.1	64.53	53.03

Table 2 Objective dialogue success and subjective dialogue quality rating by participants

	DQN-DIP	GPSARSA-DIP	GPSARSA-BCM
Success	78.6%	59.3%	75%
Rating	3.78	2.67	3.5
Turns/Task	3.25	5.04	4.17

4 Conclusion

We have presented a novel approach to training multi-domain dialogue managers using DNNs. The core idea in this method is to train a single, domain-independent policy network that can be applied to information-seeking dialogues. This is achieved through DIP [17] and as we have shown in this paper, DNNs trained with DQN [9] perform very well. We are currently exploring different DNN architectures and techniques that extend the present work and can handle large action spaces and multi-modal state spaces. Last, we plan to combine our DM with belief state tracking methods such as [19] or [6], in an effort to move towards end-to-end learning.

References

1. Casanueva I, Hain T, Christensen H, Marxer R, Green P (2015) Knowledge transfer between speakers for personalised dialogue management. In: 16th SIGDial, p 12
2. Chandramohan S, Geist M, Lefevre F, Pietquin O (2014) Co-adaptation in spoken dialogue systems. In: Natural interaction with robots, knowbots and smartphones, pp 343–353
3. Cuayáhuitl, H, Yu S, Williamson A, Carse J (2016) Deep reinforcement learning for multi-domain dialogue systems. arXiv preprint arXiv:1611.08675
4. Gašić M, Kim D, Tsiakoulis P, Young S (2015) Distributed dialogue policies for multi-domain statistical dialogue management. In: 2015 IEEE international conference on acoustics, speech and signal processing (ICASSP). IEEE, pp 5371–5375
5. Gašić M, Mrkšić N, Su P, Vandyke D, Wen T, Young S (2015) Policy committee for adaptation in multi-domain spoken dialogue systems. In: IEEE workshop on automatic speech recognition and understanding (ASRU). IEEE, pp 806–812
6. Lee S, Stent A (2016) Task lineages: dialog state tracking for flexible interaction. In: 17th annual meeting of the special interest group on discourse and dialogue, p 11
7. Lemon O, Georgila K, Henderson J (2006) Evaluating effectiveness and portability of reinforcement learned dialogue strategies with real users: the talk towninfo evaluation. In: IEEE spoken language technology workshop. IEEE, pp 178–181
8. Margolis A, Livescu K, Ostendorf M (2010) Domain adaptation with unlabeled data for dialog act tagging. In: Proceedings of the 2010 workshop on domain adaptation for natural language processing. ACL, pp 45–52
9. Mnih V, Kavukcuoglu K, Silver D, Rusu A, Veness J, Bellemare M, Graves A, Riedmiller M, Fidjeland AK, Ostrovski G et al (2015) Human-level control through deep reinforcement learning. Nature 518(7540): 529–533
10. Papangelis A, Stylianou Y (2016) Multi-domain spoken dialogue systems using domain-independent parameterisation. In: Domain adaptation for dialogue agents

11. Schaul T, Quan J, Antonoglou I, Silver D (2015) Prioritized experience replay. arXiv preprint arXiv:1511.05952
12. Sutton R, Barto A (1998) Reinforcement learning: an introduction. MIT Press
13. Taylor ME, Stone P (2009) Transfer learning for reinforcement learning domains: a survey. J Mach Learn Res 10: 1633–1685
14. Tur G, Guz U, Hakkani-Tur D (2006) Model adaptation for dialog act tagging. In: IEEE spoken language technology workshop. IEEE, pp 94–97
15. Ultes S, Rojas-Barahona L, Su P, Vandyke D, Kim D, Casanueva I, Budzianowski P, Mrkšić N, Wen T, Gašić M, Young S (2017) Pydial: a multi-domain statistical dialogue system toolkit. In: ACL 2017 Demo, Vancouver, ACL
16. Walker M, Stent A, Mairesse F, Prasad R (2007) Individual and domain adaptation in sentence planning for dialogue. J Artif Intell Res 30:413–456
17. Wang Z, Wen T, Su P, Stylianou Y (2015) Learning domain-independent dialogue policies via ontology parameterisation. In: 16th annual meeting of the SIGDial, p 412
18. Young S, Gašić M, Keizer S, Mairesse F, Schatzmann J, Thomson B, Yu K (2010) The hidden information state model: a practical framework for POMDP-based spoken dialogue management. Comput Speech Lang 24(2):150–174
19. Zhao T, Eskenazi M (2016) Towards end-to-end learning for dialog state tracking and management using deep reinforcement learning. arXiv preprint arXiv:1606.02560
20. Zhu S, Chen L, Sun K, Zheng D, Yu K (2014) Semantic parser enhancement for dialogue domain extension with little data. In: SLT workshop. IEEE, pp 336–341

An Assessment Framework for DialPort

Kyusong Lee, Tiancheng Zhao, Stefan Ultes, Lina Rojas-Barahona, Eli Pincus, David Traum and Maxine Eskenazi

Abstract Collecting a large amount of real human-computer interaction data in various domains is a cornerstone in the development of better data-driven spoken dialog systems. The DialPort project is creating a portal to collect a constant stream of real user conversational data on a variety of topics. In order to keep real users attracted to DialPort, it is crucial to develop a robust evaluation framework to monitor and maintain high performance. Different from earlier spoken dialog systems, DialPort has a heterogeneous set of spoken dialog systems gathered under one outward-looking agent. In order to access this new structure, we have identified some unique challenges that DialPort will encounter so that it can appeal to real users and have created a novel evaluation scheme that quantitatively assesses their performance in these situations. We look at assessment from the point of view of the system developer as well as that of the end user.

Keywords Dialog systems · Dialog evaluation

K. Lee (✉) · T. Zhao · M. Eskenazi
Carnegie Mellon University, Pittsburgh, PA, USA
e-mail: kyusongl@cs.cmu.edu

T. Zhao
e-mail: tianchez@cs.cmu.edu

M. Eskenazi
e-mail: max@cs.cmu.edu

S. Ultes · L. Rojas-Barahona
Cambridge University, Cambridge, UK
e-mail: stefan.ultes@eng.cam.ac.uk

L. Rojas-Barahona
e-mail: lmr46@eng.cam.ac.uk

E. Pincus · D. Traum
University of Southern California, Los Angeles, CA, USA
e-mail: pincus@ict.usc.edu

D. Traum
e-mail: traum@ict.usc.edu

M. Eskenazi et al. (eds.), *Advanced Social Interaction with Agents*, Lecture Notes in Electrical Engineering 510, https://doi.org/10.1007/978-3-319-92108-2_10

1 Introduction

Data-driven methods have become increasingly popular in developing better spoken dialog systems (SDS) due to their superior performance and scalability compared to manual handcrafting [6, 8]. DialPort [1, 9–11] is providing a new solution for rapidly collecting conversational data. The goal of DialPort is to combine a large number of dialog systems that have diverse functionality in order to attract a group of stable real users and to maintain those users' interest. In order to ensure that the real users are attracted to DialPort over the long term, we need a principled framework for monitoring and improving its performance. In this manner, at any time we can at any time have a snapshot of system performance and quickly make changes so that we do not lose our users.

Past SDS assessment paradigms [2, 7] may not be directly applicable to DialPort because it groups multiple remote agents that are heterogeneous in nature. Thus this paper proposes a novel assessment scheme based on the PARADISE framework [7] which was designed to measure SDSs' user satisfaction. Specifically, our scheme assesses DialPort according to its ultimate goals: collecting large amounts of data, and satisfying the real users' needs. For this, the assessment must reflect *how* the portal achieves the former goal: the smooth character of the conversation, response delay and performance.

In order to verify the effectiveness of the proposed assessment paradigm, we conducted a real-user study using the DialPort portal agent, which transfers control of the dialog, according to user needs, to five remote agents: (1) a weather information system from CMU (using NOAA[1]), (2) a restaurant information system from CMU (using YELP[2]), (3) another restaurant system from Cambridge University [3], (4) a word guessing game agent from the University of Southern California (USC) [4] and (5) a chat-bot from CMU and POSTECH. The CMU systems and the portal are on-site and the others are connected from off-site locations. We quantitatively assessed the performance of the DialPort portal in both task success rate, dialog management efficiency and speed of response. Finally, we show that our evaluation framework provides robust measures for DialPort.

2 Evaluation Framework of DialPort

PARADISE [7] measures the user satisfaction (US) in terms of the *success* and *cost*, where success measures *what* a system is supposed to accomplish and cost measures *how* a system achieves its goal. Also, in order to achieve highest US, a system should maximize success and minimize cost. Based on this formulation, we lay out the structure of the measure of US for DialPort in Fig. 1. The following sections explain the elements included in our measure of US in detail.

[1]http://www.noaa.gov/.

[2]https://www.yelp.com/developers/documentation/v2/overview.

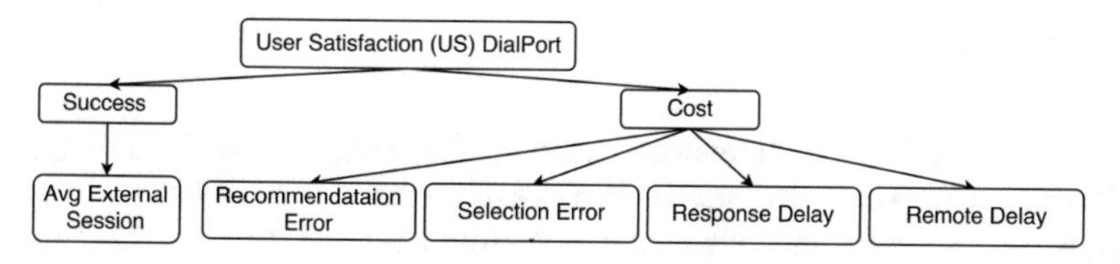

Fig. 1 The decomposition of user satisfaction for DialPort

2.1 Success of DialPort

The goal of DialPort is to collect large amounts of spoken dialog system data from real users, including both successful and failed dialogs. For clarity, we refer to a dialog with a remote agent as *external sessions* and the whole conversation from the moment the user enters the Portal to the end when they leave as a *portal session* (a portal session is composed of multiple external sessions). Therefore, the success of DialPort could be measured in terms of the average number of external sessions per portal session (avgExtSess). However in order to keep real users attracted to DialPort and in turn create more data, it is essential to maintain good conversational flow and successful performance. Therefore, we also must minimize the cost, which is defined below.

2.2 Cost

The cost, in our context, can be defined as:

- Response Delay of DialPort (D): responding to users with a minimal delay is crucial for an SDS [5] in order to maintain the pace of the interaction and to avoid barge-ins and system interruptions of the user. We measure the average delay in milliseconds from the point at which the user finishes speaking to the beginning of the next system response.
- Selection Error (S): refers to the situation when the Portal agent, which must decide which agent to connect to the user, switches to a suboptimal external agent. We measure this in terms of selection error rate.
- Recommendation Error (R): refers to the case where users do not agree with the Portal agent's recommendation. We measure this in terms of recommendation error rate.
- Remote Delay (RD): Although the success of DialPort could depend on the number of dialogs it gathers, the quality of the external agents greatly impact user satisfaction. Therefore, we include the average response delay with remote agents as a part of the cost, in order to avoid disruptions such as barge-ins and system interruptions.

2.3 Overall Performance

The original PARADISE framework learns a linear regression [7] to weight the importance of each cost and success in order to predict the users' subjective scores. In this work, we assume each element is uniformly weighted. Thus we first normalize each input and compute the overall US of DialPort by:

$$US_{meta} = \mathcal{N}(avgExtSess) - (\mathcal{N}(S) + \mathcal{N}(R) + \mathcal{N}(D) + \mathcal{N}(RD)) \quad (1)$$

where $\mathcal{N}$ stands for z normalization to standardize the input into zero mean and unit variance, to ensure measurements with different scales can be combined together.

3 Evaluation

We assessed DialPort as an agent that seamlessly changes domains. We analyzed log data of the current version of DialPort. We gathered 119 dialogs from a group of 10 CMU Language technologies Institute students. The maximum number of dialogs per any one user was 11 and the minimum number of dialogs was 1. On average a dialog with DialPort lasted 10.97 turns. Two experts manually tagged the selection error and the recommendation error. Other measures were automatically obtained from the log data such as the number of sessions and response delay.

3.1 Evaluation of External Remote Agents

First, we show statistics such as the number of external sessions and response delay of groups of external agents (Table 1). We looked at: all external agents combined, all

Table 1 Performance of external agents

	Portal agent	All external agents	Two on-site (w/o chatbot)	Three on-site (plus chatbot)	Two off-site
# of external sessions	–	614	135	552	62
Avg # of turn/external session	–	2.35	2.54	1.37	11.01
Avg resp delay (ms)	457.80	683.54	518.97	683.39	683.89
Std resp delay (ms)	456.86	529.74	333.72	488.56	599.19

three on-site agents, both off-site agents, and two CMU agents excluding the chatbot. We observed that many user requests are directed to the chatbot. Moreover, 28.8% of chatbot utterances were non-understanding recovery turns, such as "can you please rephrase that?", "sorry I didn't catch that". The remaining 71.2% of the chatbot turns were question-answering and chit-chat. Based on the log data collected, we observed that users tend to talk to the chatbot after finishing a conversation with one of the task-oriented external agents. After a few turns of interactions with the chatbot, users usually initiate new external sessions with other task-oriented external agents. This shows the importance of handling out-of-domain utterances gracefully to maintain the flow of the dialog.

We also note that the external sessions with off-site agents are longer than those with the on-site ones. While the chatbot is responsible for a lot of this difference, we note that users can carry on longer conversations with the external agents since our goal is to create a data flow for these systems.

There are two main causes of response delay: a network delay and a computation delay. The servers for the Portal agent and the on-site remote agents are located in Pittsburgh, so the network delay between them and the other systems was small. Off-site systems have a longer network delay. Table 1 shows that the total response delay of on-site systems is longer than the delay of off-site systems. This is because the chatbot accesses a very large database, which introduces a significant amount of computation time. By excluding the chatbot in our calculations, we see that the response delay of on-site systems is 24.1% ($P - value < 0.001$) faster than the off-site systems.

3.2 *Evaluation of the Portal Agent*

We then separately assessed the Portal agent, that is, the agent that the user interacts with that decides which SDS to connect to the user. We assess the Portal agent for (1) recommendation error (2) selection error. Recommendation errors were labeled for each system utterance with the recommendation intent with binary label {0, 1}, where 1 means correct and 0 means incorrect (Table 2).

A recommendation score gets Label 1 if the users agree with the recommendation, otherwise it is 0. For example, when a system said "Would you like to know about next weekend's weather?", a positive reward would be received if the next user utterance is "yes", "okay" or "what is the weather in Pittsburgh?". Otherwise, the

Table 2 Performance of the Portal agent. Since there is only one version of the Portal agent in this study, we cannot z-normalize the scores. AvgExtSess is number of external systems accessed in one portal session

	AvgExtSess	Select error	Recommend error
Master agent	4.32	22.6%	37.15%

Fig. 2 The external system confusion matrix for the Portal

	System A	System B	System C	System D	System E
System A	0.958	0.000	0.000	0.000	0.042
System B	0.071	0.857	0.000	0.000	0.071
System C	0.000	0.000	0.900	0.000	0.100
System D	0.000	0.059	0.000	0.824	0.118
System E	0.140	0.053	0.009	0.000	0.798

system would get the label 0. The result shows that 62.85% of the requests were successfully recommended. Selection errors were labeled for each system utterance that switches to a new agent. If the transition is correct, it gets label 1, otherwise 0. For example, if a user said Who is the president of South Korea? the Portal agent should select the chatbot. Any other selection will result in label 0. 78.40% were successfully sent to the appropriate agents. Figure 2 shows that the most frequent error was due to the Portal agent selecting a weather or restaurant agent when a location entity was detected in chit-chat utterances.

4 Conclusion

We have proposed a framework for the assessment of DialPort and evaluated the Portal agent and its access to the external system based on our collected annotated conversational data. We have identified unique challenges that DialPort faces and have provided a comprehensive evaluation framework that covers the performance of response delay, agent transition, and recommendation strategy for the portal. In future work, we plan to develop data-driven models to automatically predict the success and cost of the Portal and remote agents of DialPort, which suggests a promising research direction.

References

1. Lee K, Zhao T, Du Y, Cai E, Lu A, Pincus E, Traum D, Ultes S, Barahona LMR, Gasic M et al (2017) Dialport, gone live: an update after a year of development. In: Proceedings of the 18th annual SIGdial meeting on discourse and dialogue, pp 170–173
2. Liu CW, Lowe R, Serban IV, Noseworthy M, Charlin L, Pineau J (2016) How not to evaluate your dialogue system: an empirical study of unsupervised evaluation metrics for dialogue response generation. arXiv preprint arXiv:1603.08023
3. Mrkšić N, Séaghdha DO, Thomson B, Gašić M, Su PH, Vandyke D, Wen TH, Young S (2015) Multi-domain dialog state tracking using recurrent neural networks. arXiv preprint arXiv:1506.07190
4. Pincus E, Traum D (2016) Towards automatic identification of effective clues for team word-guessing games. In: Proceedings of the language resources and evaluation conference (LREC). European Language Resources Association, Portoro, Slovenia, pp 2741–2747

5. Raux A, Eskenazi M (2009) A finite-state turn-taking model for spoken dialog systems. In: Proceedings of human language technologies: the 2009 annual conference of the north american chapter of the association for computational linguistics. Association for Computational Linguistics, pp 629–637
6. Vinyals O, Le Q (2015) A neural conversational model. arXiv preprint arXiv:1506.05869
7. Walker MA, Litman DJ, Kamm CA, Abella A (1997) Paradise: a framework for evaluating spoken dialogue agents. In: Proceedings of the eighth conference on European chapter of the association for computational linguistics. Association for Computational Linguistics, pp 271–280
8. Williams JD, Young S (2007) Partially observable markov decision processes for spoken dialog systems. Comput Speech Lang 21(2):393–422
9. Zhao T, Eskenazi M, Lee K (2016) Dialport: a general framework for aggregating dialog systems. EMNLP 2016:32
10. Zhao T, Lee K, Eskenazi M (2016) Dialport: connecting the spoken dialog research community to real user data. In: 2016 IEEE workshop on spoken language technology
11. Zhao T, Lee K, Eskenazi M (2016) The dialport portal: grouping diverse types of spoken dialog systems. In: Workshop on Chatbots and conversational agents

Part III
Human-Robot Interaction

Towards Metrics of Evaluation of Pepper Robot as a Social Companion for the Elderly

Lucile Bechade, Guillaume Dubuisson-Duplessis, Gabrielle Pittaro, Mélanie Garcia and Laurence Devillers

Abstract For the design of socially acceptable robots, field studies in Human-Robot Interaction are necessary. Constructing dialogue benchmarks can have a meaning only if researchers take into account the evaluation of robot, human, and their interaction. This paper describes a study aiming at finding an objective evaluation procedure of the dialogue with a social robot. The goal is to build an empathic robot (JOKER project) and it focuses on elderly people, the end-users expected by ROMEO2 project. The authors carried out three experimental sessions. The first time, the robot was NAO, and it was with a Wizard of Oz (emotions were entered manually by experimenters as inputs to the program). The other times, the robot was Pepper, and it was totally autonomous (automatic detection of emotions and decision according to). Each interaction involved various scenarios dealing with emotion recognition, humor, negotiation and cultural quiz. The paper details the system functioning, the scenarios and the evaluation of the experiments.

Keywords Human-Robot Interaction · Data collection · Evaluation
Elderly end-users · Metrics

1 Introduction

Currently, several research teams work on projects for the elderly self-sufficiency [4, 11, 12], particularly on conversational agents design [1, 15]. To build a coherent and engaging conversational agent, social dialogue is essential. An autonomous robot with clever perceptual analysis is more engaging for the user. Furthermore, it may lead to personalize relationship with the user [3]. Empathy may help a lot in the analysis

L. Bechade (✉) · G. Dubuisson-Duplessis · G. Pittaro · M. Garcia · L. Devillers
LIMSI-CNRS, Université Paris-Sud, 91405 Orsay Cedex, France
e-mail: lucile.bechade@limsi.fr

L. Devillers
Université Paris-Sorbonne, 75005 Paris, France
e-mail: laurence.devillers@limsi.fr

M. Eskenazi et al. (eds.), *Advanced Social Interaction with Agents*, Lecture Notes in Electrical Engineering 510, https://doi.org/10.1007/978-3-319-92108-2_11

and decision of answer tasks. The purpose of JOKER (JOKe and Emphathy of a Robot) project is to give a robot such a capability, as well as humor. Besides, regarding human-robot dialogue, neither a clear common framework on social dialogue nor a procedure of evaluation exist. Aly et al. [2] tried to find metrics so as to better evaluate Human-Robot Interaction (HRI). In this paper, the authors introduce a study as part of the ROMEO2 project. They explain the context, the system description, the proceedings of three HRI experimental sessions carried out with elderly people. They aim at evaluating objectively their multi-modal system in order to build a real personal robot assistant for elderly people.

2 Related Work

In Human-Robot Interaction, researches have focused on elderly users, and on the robot's ability to help the participant: to stand [18], to catch something [9] or to walk [17]. To maintain user engagement and increase acceptability of robot, social dialogue is crucial. Indeed, regarding media interaction, Reeves and Nass [14] emphasize that people react to media as if they were social actors. Several works address the issue of evaluating Human-Robot spoken interactions in a social context by considering the engagement of the human participant [8].

For dialogue systems, the theoretical framework PARADISE [16] evaluate dialogue systems from combined measures of the system's performance. This evaluation is performed from the point of view of the performance required to achieve user's satisfaction. The main goal of the system should be to maximize user satisfaction. Therefore, evaluating a system according to the PARADISE framework requires to define the objectives of the system and to characterize them. In human-robot interaction, metrics of evaluation are also proposed. In a context where the robot has no specified task, where the user is free to move and evolve spontaneously around the robot, and when no satisfaction questionnaire can be performed, a global evaluation is particularly difficult to achieve.Dautenhahn and Werry [5] provides a descriptive and quantitative analysis of low-level behaviors to evaluate the interaction.

In assistive and social robotics, experiments with potential end-users provide a valuable feedback about researchers' expectations, and reliable data for the design of socially acceptable robots. Moreover, user feedback may help to improve the evaluation and the development of dialogue system. In that regard, [19] worked on an evaluation plan based on incremental stages corresponding to the improvement of a dialogue system according to user feedback.

3 Context

3.1 JOKER Project

Social interactions require social intelligence and understanding: anticipating the mental state of another person may help to deal with new circumstances. JOKER researchers investigate humor in human-machine interaction. Humor can trigger surprise, amusement, or irritation if it does not match the user's expectations. They also explore two social behaviors: expressing empathy and exchanging chat with the interlocutor as a way to build a deeper relationship.The project gathers 6 international partner laboratories. LIMSI involvement is on affective and social dimensions in spoken interaction, emotion and affect bursts detection, user models, Human-Robot Interaction, dialogue, generation.

3.2 ROMEO2 Project

Aldebaran launched ROMEO2 project with the objective to build and develop a personal robot companion for the elderly [10]. On the one hand, the ROMEO robot will be capable of moving, take items and give simple information to the user. On the other hand, it will be able to think, to reason, in order to detect extraordinary situations, to decide, to adapt its answers. The multi-modality goal with multi-sensory perception and cognitive interaction (reasoning, planning, learning mechanisms) makes this project unique. Indeed, ROMEO will be able to assist old people and to answer the best it can to their requests using both a predefined database and a learned database through its experiences. ROMEO will learn from its everyday life with the user and will be able to adapt and personalize its behavior. Furthermore, its capacity to detect emotions from the user will make its process decide the best behavior to adopt during a dialog. LIMSI is on emotion recognition, multimodal speaker identification, speech comprehension and social interaction.

3.3 LUSAGE Living Lab

LUSAGE Living Lab [13] at the Broca hospital, Paris, welcomed the experiments, under the supervision of the gerontology service. Regularly, the Living Lab organizes workshops in the "Café Multimédia" project. Most of elderly people are badly aware of digital technologies, becoming more and more socially isolated. The goal of the project is to bridge that divide. The participants can discover the Information and Communication Technologies, discuss them, and meet designers and researchers. The authors of this study took part in the activities of the Lab: workshop on social robotics for health-care and everyday life. On these occasions, they offered the par-

Fig. 1 Pictures taken during interactions between: an elderly and Nao (on the left), an elderly and Pepper (on the right)

ticipants to interact with two robots (Nao and Pepper). After each experiment, the researchers discussed (individually and in group) with them, about the experiment, but also about their opinions on social and assistive robotics in general (Fig. 1).

4 Description of the Experiment

4.1 Experimental Process

First of all, the researchers gave to participants a general and collective explanation of the experiment: the aims of the researchers, the type of interaction and the nature of the robot they were going to meet. They carried out three experimental sessions within the same context. In the first session, the Humor system was a Wizard of Oz (totally operated by a human experimenter) with Nao. The experimenter provided a part of the inputs manually (emotion detection). In the second and third ones, all the system was autonomous. The authors used Pepper. Each participant of the first and second session interacted with the robot only once while those of the third between one and four times. The comparison between the experiments is shown in Fig. 2.

	1st Experiment	2nd Experiment	3rd Experiment
Number of participants	12	8	22
(whose elderly)	12	4	16
(per person)	1	1	1 to 4
Robot	Nao	Pepper	Pepper
Type of system	Wizard of Oz	Autonomous	Autonomous
Scenarios	Humor, Negotiation	Emotion game, Humor, Negotiation, Quiz	Emotion game, Humor, Negotiation, Quiz

Fig. 2 Table comparing the experiments

4.2 Description of the System

The system during the two last experimental sessions was totally autonomous. Each module of analysis of linguistic and paralinguistic could communicate through a multi-modal platform. The multi-modal platform enables to make the link between paralinguistic module and decision process while running the scenarios. Several levels of abstraction constitute the system, that allow to build a dialogue system on top of the crossbar architecture. From the lowest level to the highest level, the system uses:

- **The Event**: technical messages exchanged by crossbar components, can be asynchronous or synchronous
- **The Contribution**: a dialogic contribution within a dialogue turn which aggregates data from the input modules (e.g., linguistics, paralinguistics)
- **The Expectations**: expectations defined at the scenario level (e.g., an emotionally positive contribution, a given word, a silence, etc.)

After most Pepper interventions, the multi-modal platform requests a contribution from the user. It takes into account : a minimal time to wait for a contribution (if something is said, Pepper will stop listening after that time) and a maximal time to wait for a contribution (if nothing is said at all, Pepper will stop listening at that point). The multi-modal platform works according to the architecture shown in Fig. 3.

The paralinguistic system features an emotion detection module based on audio [7]. The audio signal is cut into segments between 200 and 1600 milliseconds in length. Each segment contains or not a detected emotion. The robot takes into account the majority emotion during a speech turn. The emotion recognition module works with a linear Support Vector Machines (SVM) with data normalization and acoustic descriptors such as acoustic parameters in the frequency domain (e.g. fundamental frequency F0), in the amplitude domain (e.g. energy), in the time domain (e.g. rhythm) and in the spectral domain (e.g. spectral envelop or energy per spectral bands).

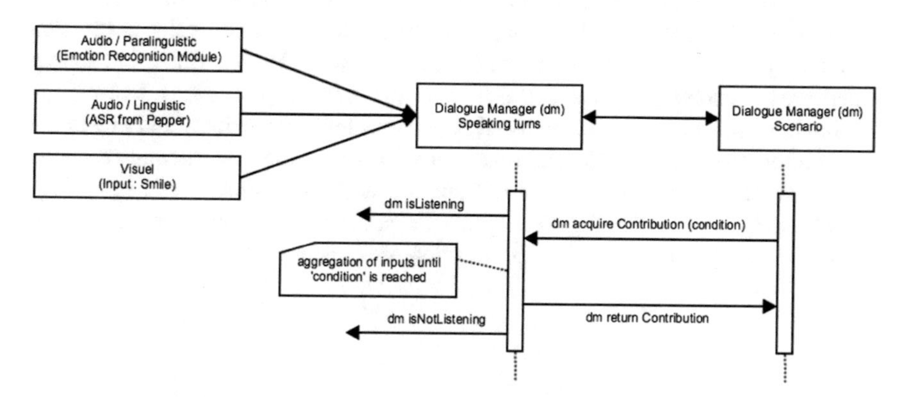

Fig. 3 Multi-modal platform system for social dialogue with Pepper

4.3 Interaction Scenarios

The tested scenarios relate to possible daily life interactions between the robot companion and an elderly person.

- **Emotions**: A first scenario consists in asking the user to mimic chosen emotions while speaking. Pepper asked the user to imitate the four emotions it can detect: Joy, Sadness, Anger, Neutral. If the speech from the user contains the requested emotion, the robot goes on following the rest of the scenario. If it does not, the robot says which major emotion it detects and asks again the user to mimic (over 3 times, the robot stop asking and proceeds to the next question). This scenario enables to evaluate the emotion recognition system performance.
- **Jokes/Riddles**: Pepper can make several riddles and puns. For instance:
  - Question: “What is a cow making while closing its eyes?”
    Answer: “Concentrated milk!”
  - Question: “How do we call a dog without legs?”
    Answer: “We do not call it, we pick it up.”

  During the experiment, according to the emotion detection on the user, Pepper adapted its humor to the user emotion profile. Pepper may also stimulate the memory of the user by asking to repeat one joke he made. The paralinguistic module takes an important part in the humor process: it detects emotions, laugh. The behavior of the system depends on the receptiveness of the human to the humorous contributions of the robot. Positive reactions (e.g. laughter, positive comments or positive emotions) lead to more humorous contributions, whereas repeated negative reactions (e.g. sarcastic laughter, negative comments and negative emotions) drive the dialogue to a rapid end. If there is no reaction, the robot tries to change its kind of humor so as to make the user react.
- **Persuasion/Negotiation**: The robot as a companion has to take care of the user showing initiative. In this experiment, Pepper tried to convince old people to drink a glass of water. Before the simulation, the user was said to always refuse the proposition from Pepper. According to the user global reaction valence detected by the system, the robot chose a negotiation strategy: Humor (the robot makes derisive comments about itself so as to make the user accept its offer), Reason (the robot argues reasonably), Calming (the robot ensures it does not want to force the user after detecting anger). The robot calculates the global valence of reactions (positive or negative) from the user. Then, it can adapt its strategy of negotiation.
- **Cultural Quiz**: The robot makes the user listen to extract of music or movies, and asks the user to recognize the singer or actor, or the name of the song or the movie.

5 Results

5.1 Evaluating the System's Performance

The emotion recognition algorithm was built with the annotated audio corpus JEMO [6]. The performance of the emotion algorithm calculated by cross-validation is described by a F-score equal to 62,4. Old people assessed the robot had difficulties to detect their emotions correctly. To evaluate the performance of our system on old people, the authors annotated with emotion labels (anger/sadness/joy/neutral) all the segments where the robot detected an emotion. The best detection performance was one correct detection over two. The worse was one over eight. The annotators noticed that sometimes segments were too short to recognize an emotion correctly. Figure 4 shows that good detections occur mostly for segment lengths higher than one second. The corpus JEMO contains voices from people aged between 20 and 50 years old. The experiments participants were between 70 and 85. A higher speech rate (for set segment length) has been observed in the corpus JEMO than in the experimental corpus. This raises the problem on speech velocity variability. Therefore, it is necessary to take into account this feature while learning emotion detection algorithm, so to take into account the feature "age" implicitly. Thanks to the data, the researchers will build a new emotion detection algorithm adapted to the elderly in future work.

5.2 Engagement Metrics

User engagement is a precious piece of information. Being able to detect it, the robot could adapt its social behavior. Thus, engagement would be an essential feature to build a dialog adaptive algorithm. Indeed, the authors wondered if the more a user is involved in the interaction, the more it talks. Moreover, if a user reacts relatively quickly, does it mean it is sensitive (positively or negatively) to what the robot is

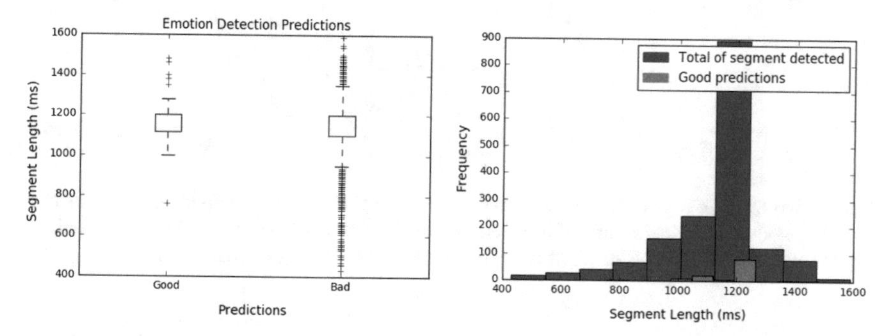

Fig. 4 Performance of the emotion detection on the elderly: type of prediction boxplot and histogram according to segment lengths (in milliseconds). Annotation errors may put a bias

saying? After each interaction, the user had to fill a questionnaire. This was about the user's feelings during the interaction and global view on the interaction and on robots. In this section, answers related to engagement are studied: did you feel involved in the interaction? Did you feel self-confident? The authors assumption is that engagement can be seen in three metrics: reaction time, silence time and speaking time during one speech turn of the user. To start a validation on that hypothesis, the authors compare these metrics to reliable information on engagement.

Figure 5 aims at highlighting links between speaking time and user engagement, and between speaking time and user self-confidence. Correlations between means of these features are respectively 0.866 and 0.881. Mean curves follow similar tendencies. The authors remind about the different sample sizes for the feature "Number of Interactions": 16 for the level one, 10 for the level two, 7 for the level three, 2 for the level four. Therefore, the hypothesis about speaking time as an engagement metric has to be validated with bigger samples in the future. The engagement of the user may be explained by two variables: the understandability of the task requested (the user understands what the robot expects him to do so he can act spontaneously), and the attractiveness of the scenario (the scenario inspires the user who can answer faster than if it does not). Adding the metric success to the Emotion Challenge and to the Quiz may also help to distinguish those who played the game from those who did not. Furthermore in a next work, a metric about user engagement during the humor scenario will be studied. Figure 6 may also show interesting link between silence time and robot dominance evaluation. If silence time represented a reliable metric of robot dominance, robot could adjust its behavior appearing more humble for the user. Thanks to this short longitudinal session and other experiments planned next, the authors expect to build new strategies on engagement of the user during Human-Machine Interaction adapted to the elderly.

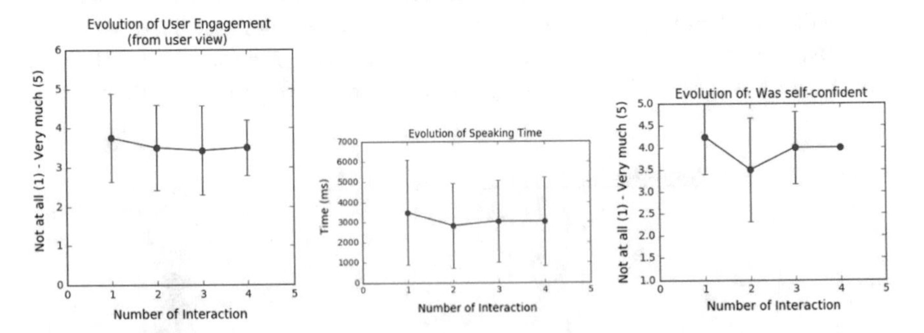

Fig. 5 Evolution of user engagement (self-evaluation), speaking time (individual means of time speaking at each speech turn) and of user self-confidence according to the number of interactions. Graphs represent means and standard-deviation. Mean curves follow similar tendencies

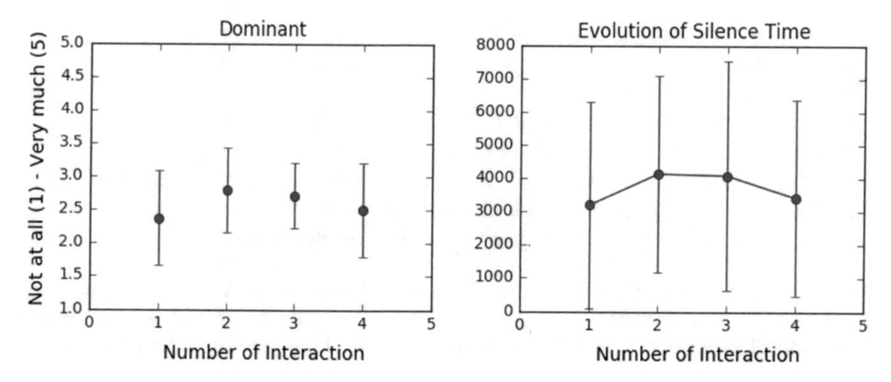

Fig. 6 Evolution of robot dominance (user evaluation) and silence time according to the number of interactions (individual means of silence time at each speech turn). Graphs represent means and standard-deviation. Mean curves look to follow similar tendencies

5.3 *Interaction Appraisal*

The researchers took recommendations from the users through questionnaire about their opinion on the robot operating, their feelings during the interaction and their thoughts about such a technology. In the first experimental session, the experimenters entered manually each user's emotional expressions. The users liked interacting with the robot. Thus, the researchers could study the acceptability of robot as a vector of communication for the elderly. However, users were different between each experimental session, and sample size was the smallest in the first one. It is not possible to draw conclusions about the impact of the autonomy of robot.

Regarding the second and third experimental sessions, the robot detected correctly most of the elderly moods (positive or negative attitude in reactions), thanks to the summarization of emotions detected in speech turns. Thus, the robot could adapt its answers correctly. Sometimes, it failed at recognizing the speech from the user and then gave incoherent answers. Figure 7 shows an example of an interaction which

P: "It's been a long time since you drank for the last time.
Take some water please."
U: "Not at all."
P: "You should drink a bit."
U: "I'm not thirsty, I only drink when I'm thirsty."
P: "So I'm going to make jokes until you accept.
Take your glass and I'll show you how I do a headstand!"
U: "It does not interest me, do your headstand alone !" *(ANGER)*
P: "Calm down, I don't do this to command you.
I only suggest you to take a glass of water, it's not that much."

Fig. 7 Example of an interaction which functioned correctly with a user: Pepper adapted its behavior correctly in relation to the user's reactions (scenario of negotiation, "P" for Pepper, "U" for User). Consequently, the user got more involved into the conversation

P: "Where does this extract come from?"
(Music Title: "Ya d'la joie", Singer: "Charles Trenet")
"Do you know who is it ?"
U: "Mmm I think it is Trenet but I'm not sure..." *(LOW VOICE)*
P: "Can you repeat please ? I didn't understand well."
U: "Maybe Charles Trenet, I don't know..." *(LOW VOICE and HESITATION)*
P: "You look wrong, but you look like a good sport!
It was an extract from "Ya d'la joie", sung by Charles Trenet."

Fig. 8 Example of an interaction which worked badly: Pepper did not detect correctly the good answer from the user (scenario of quiz, "P" for Pepper, "U" for User). Consequently, the user doubted about the quality of the conversation and turned unwilling to dialog with the robot

functioned correctly with a user while Fig. 8 shows one of an interaction which worked badly.

What the researchers drew from these experiments is positive. Participants reported on the socio-affective states they felt during the experiment in self-report questionnaire. Participants feelings were globally positive: participants assessed to have more positive affective states than negative ones during the interaction. Figure 9 shows the evolution of enjoyment interacting with the robot according to the number of interactions (third experimental session). The slight decrease between the first and the second time may be explained by a feeling of lassitude and habituation: users do not have the same surprise than in their first interaction.

The size of the sample does not allow statistically significant results but it gives an interesting overview on the interest of the adaptive behavior of the robot. As for the preference between each scenario, the elderly preferred the emotion game and the quizz to a great extent. Those correspond to scenarios in which the robot clearly asks the user to say something. The users talked much more during the emotion game. Moreover, the quizz made them react unequally. Some of them talked a lot during the quizz, others did not dare to (maybe scared of being wrong). Moreover,

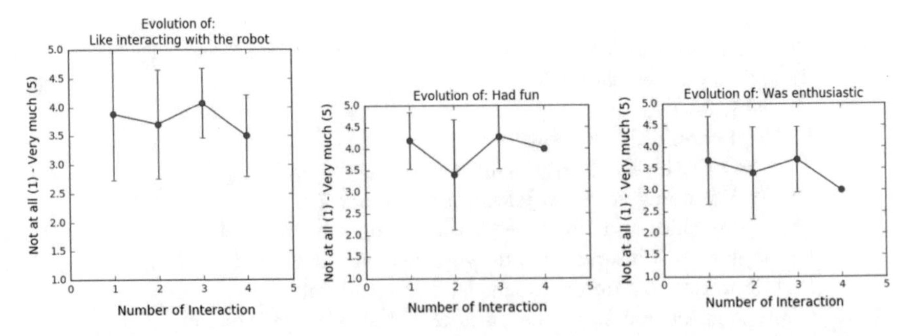

Fig. 9 Third experimental session—answers to enjoyment questions ("did you like interacting with the robot?", "did you have fun?", "did you feel enthusiastic?") according to the number of interaction. The slight decrease between the first and second time interacting with the robot may be explained by user habituation

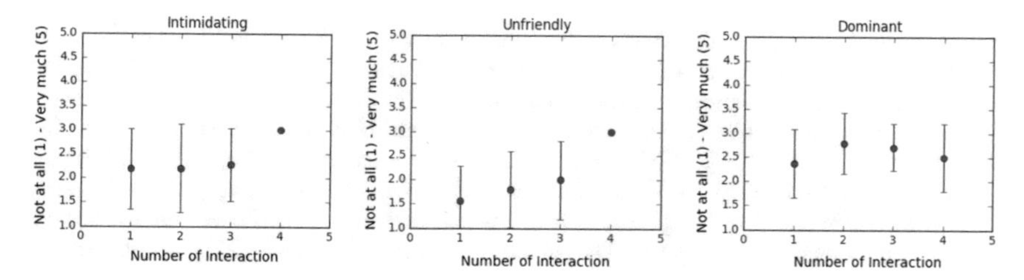

Fig. 10 Third experimental session—answers about robot behavior (the robot was comforting/intimidating, friendly/unfriendly, humble/dominant) according to the number of interactions. This shows the participant view variability about the concept of "normal" robot

this longitudinal study will help providing more data to find new benchmarks specific to interactions between robot and elderly people. These will be taken into account to evaluate the system. Nevertheless, a robot which perfectly fits everyone is a hard to reach ideal. Indeed, Fig. 10 shows the difficulty to have a robot whose all the elderly finds normal in its behaviors. It shows the diversity between user evaluation of robot behavior during the interaction. That result may also be due to the adaptive behavior and different strategies it used according to the user reactions detected. Thus, there is a double variability: that of the user personalities, that of the robot choices on behavior strategies.

6 Conclusion and Discussion

To build functional and socially acceptable conversational agents, having benchmarks is primordial. What the authors are trying to do is to find metrics and benchmarks by collecting data from the end-users, in order to evaluate objectively their system, and to go on improving it. In the case of ROMEO2 project, these metrics may be proper to the elderly. The authors are wondering how to make it ethically adapted, efficient and understandable, useful and easy-to-use (from the elderly people perspective) to make them more easily use the technology. The authors started collecting data first with a WoZ system with Nao, next with an autonomous system and the Pepper robot. The last experimental session allowed them to study the evolution of user reactions according to the individual number of interaction with the robot. The experiments at Broca hospital emphasize main issues: the user concerns about data safety and the adaptation of the robot to the user. In this second point, it is necessary to take into account the age of the user. The robot has to adapt its vocabulary and its behavior according to the user's age. It also has to change its speech velocity and its way to detect emotions (according to the speed of the user speaking).

The system will be improved using data collected at the experiments described and more tests will be done at the Broca Hospital in collaboration with health-care workers about the acceptability of such a technology.

Acknowledgements The authors wish to thank again Pr. Anne-Sophie Rigaud, head of the gerontology department at the Broca hospital, and her team, for providing access to LUSAGE Living Lab facilities, hence allowing the authors to carry out this study. The ROMEO2 Project has been financed by Bpifrance and labeled by CAP DIGITAL competitive center (Paris Region). The JOKER Project has been sponsored by ERA-Net CHIST-ERA (http://www.chistera.eu/), the Agence Nationale pour la Recherche (ANR, France), the Fonds National de la Recherche Scientifique (FNRS, Belgium), the The Scientific and Technological Research Council of Turkey (TUBITAK, Turkey) and the Irish Research Council (IRC, Ireland). The authors would also like to show their gratitude to all the participants of the experiment.

References

1. Agrigoroaie R, Ferland F, Tapus A (2016) The enrichme project: lessons learnt from a first interaction with the elderly. In: ICSR
2. Aly A, Griffiths S, Stramandinoli F (2016) Metrics and benchmarks in human–robot interaction: recent advances in cognitive robotics. Cogn Syst Res
3. Aly A, Tapus A (2013) A model for synthesizing a combined verbal and nonverbal behavior based on personality traits in human–robot interaction. In: HRI
4. Cross dissemination agreement and implementation with Mobile Age Project H2020
5. Dautenhahn K, Werry I (2002) A quantitative technique for analysing robot-human interactions. In: IEEE/RSJ international conference on intelligent robots and systems, vol 2, pp 1132–1138. https://doi.org/10.1109/IRDS.2002.1043883
6. Devillers L, Tahon M, Sehili M, Delaborde A (2014) Détection des états affectifs lors d'interactions parlées: robustesse des indices non verbaux. TAL 55(2)
7. Devillers L, Tahon M, Sehili MA, Delaborde A (2015) Inference of human beings' emotional states from speech in human–robot interactions. Int J Soc Robot:1–13
8. Dubuisson Duplessis G, Devillers L (2015) Towards the consideration od dialogue activities in engagement measures for human-robot social interaction. In: International conference on intelligent robots and systems, designing and evaluating social robots for public settings workshop, pp 19–24
9. Kumar S, Rajasekar P, Mandharasalam T, Vignesh S (2013) Handicapped assisting robot. In: International conference on current trends in engineering and technology (ICCTET)
10. Pandey AK, Gelin R, Alami R, Vitry R, Buendia A, Meertens R, Chetouani M, Devillers L, Tahon M, Filliat D, Grenier Y, Maazaoui M, Kheddar A, Lerasle F, Duval LF (2014) Ethical considerations and feedback from social human–robot interaction with elderly people. In: AIC
11. Panou M, Bekiaris E, Touliou K, Cabrera M (2015) Use cases for optimising services promoting autonomous mobility of elderly with cognitive impairments. In: TRANSED
12. Panou M, Cabrera M, Bekiaris E, Touliou K (2015) Ict services for prolonging independent living ofelderly with cognitive impairments. In: AAATE
13. Pino M, Benveniste S, Picard R, Rigaud A (2017) User driven innovation for dementia care in France: the LUSAGE living lab case study. Interdisc Stud J 3:1–18
14. Reeves B, Nass C (1996) The media equation: how people treat computers, television, and new media like real people and places. Cambridge University Press
15. Tobis S, Cylkowska-Nowak M, Wieczorowska-Tobis K, Pawlaczyk M, Suwalska A (2017) Occupational therapy students'perceptions of the role of robots in the care for older people living in the community. Occup Ther Int
16. Walker MA, Litman DJ, Kamm CA, Abella A (1997) Paradise: a framework for evaluating spoken dialogue agents. In: Proceedings of the eighth conference on European chapter of the association for computational linguistics, EACL '97, pp 271–280. Association for computational linguistics, Stroudsburg, PA, USA. https://doi.org/10.3115/979617.979652

17. Wei X, Zhang X, Wang Y (2012) Research on a detection and recognition method of tactile-slip sensation used to control the elderly-assistant & walking-assistant robot. In: IEE (ed.) International conference on automation science and engineering (CASE)
18. Xiong G, Gong J, Zhuang T, Zhao T, Liu D, Chen X (2007) Development of assistant robot with standing-up devices for paraplegic patients and elderly people. In: IEE (ed.) International conference on complex medical engineering (CME)
19. Van der Zwaan JM, Dignum VJVdZ, Dignum V, Jonker C (2012) A bdi dialogue agent for social support: specification and evaluation method. In: International foundation for autonomous agents and multiagent systems (IFAAMAS)

A Social Companion and Conversational Partner for the Elderly

Juliana Miehle, Ilker Bagci, Wolfgang Minker and Stefan Ultes

Abstract In this work, we present the development and evaluation of a social companion and conversational partner for the specific user group of elderly persons. With the aim of designing a user-adaptive system, we respond to the desires of the elderly which have been identified during various interviews and create a companion that talks and listens to the elderly users. Moreover, we have conducted a user study with a small group of retired seniors living at home or in a nursing home. The results show that our companion and its dialogue were perceived very positively and that a social companion and conversational partner is indeed in demand by lonely seniors.

Keywords User study · Spoken dialogue · User-adaptation

1 Introduction

For humans, speech is the most important and most natural way to communicate. Therefore, scientists and engineers create methods and systems that enable not only interpersonal communication but also interaction with machines through natural spoken language. Today, we are able to communicate with various computer applications via speech. However, most of the currently used Spoken Dialogue Systems feel unnatural to humans, users are dissatisfied and dialogues are unsuccessful [6]. Therefore, current research focuses on user-adaptive Spoken Dialogue Systems (e.g. [7, 8]). In this work, we address the specific user group of the elderly.

J. Miehle (✉) · I. Bagci · W. Minker
Institute of Communications Engineering, Ulm University, Ulm, Germany
e-mail: juliana.miehle@uni-ulm.de

I. Bagci
e-mail: ilker.bagci@uni-ulm.de

W. Minker
e-mail: wolfgang.minker@uni-ulm.de

S. Ultes
Department of Engineering, University of Cambridge, Cambridge, UK
e-mail: su259@cam.ac.uk

M. Eskenazi et al. (eds.), *Advanced Social Interaction with Agents*, Lecture Notes in Electrical Engineering 510, https://doi.org/10.1007/978-3-319-92108-2_12

In recent years, assistive technologies in the area of elderly care have been a rapidly increasing field of research and development and various socially assistive companions for the elderly have been presented. The spectrum of functionalities and services reaches from intelligent reminding [5, 9], information provision and help in carrying out everyday tasks [10], via exercise advice [2] and guidance through the environments of the elderly [9], up to cognitive stimulation, mobile video-telephony with relatives or caregivers and the autonomous detection of dangerous situations like falls and their evaluation via mobile telepresence [5]. The effects and the effectiveness of socially assistive robots in elderly care have been reviewed, showing that there is a potential for the use of robot systems in elderly care [1]. Moreover, the acceptance of healthcare robots for elderly users is investigated in [3].

These findings show that there is a potential for an embodied social companion for elderly users. Furthermore, the importance of considering the perspectives from a range of stakeholders, such as the elderly person, his or her family and medical staff, and of carefully assessing their expectations and needs has been shown [3]. Therefore, the aim of our work is to develop a social companion and conversational partner for the elderly based on both the desires and wishes of seniors affected by problems of loneliness and the need for care and on the assessments by their caregivers.

Looking into this topic from a different perspective, Walters et al. [11] conducted a theatre-based human-robot interaction study. Elderly residents and caregivers thereby watched a theatre production of a play illustrating the functionality as well as social and ethical issues of robots when used in the elderly care domain. Interviews about their views indicate that both caregivers and residents were generally positive towards the idea of using robots in the care domain. Furthermore, the desire for care robots for additionally providing social interaction and entertainment was stressed by the residents. However, the trial participants have only been interviewed after the theatre about the robot's functionalities has been produced. In contrast, our goal is to involve the elderly and their caregivers from the very beginning of the development of our social companion and conversational partner.

The study which has been presented in [4] explores people's perceptions and attitudes towards future robot companions for the home. The results show that a large proportion of participants were in favour of a robot companion acting as an assistant, machine or servant. Only some of them wished a robot companion to be a friend. However, most of the participants were students or academic staff aged between 26 and 45 who talked about their future, trying to imagine themselves in the situation of an elderly person, which of course differs from the situation of directly asking the elderly. In contrast, the starting point of our work has been to talk to seniors affected by problems of loneliness and the need for care as well as with their caregivers. We have identified the desires of the elderly during various interviews at a nursing home and designed a prototype of a social companion for the elderly in accordance with them. Afterwards, we have conducted a user study with a small group of retired seniors living at home or in a nursing home.

The structure of the chapter is as follows: In Sect. 2, we introduce our specific user group of the elderly. Subsequently, the development of our social companion and the results of our study are presented in Sect. 3, before concluding in Sect. 4.

2 Requirement Analysis

With the intention of getting a first insight in the specific needs and requests of ageing adults, we have carried out discussions with the residents of a nursing home in the south of Germany as well as with the social managers and caregivers. With these groups, we have discussed about the aims and needs of the elderly. The major topic among our interlocutors was the loneliness of the elderly living in a nursing home. The fact that those people feel lonely can be easily explained. Nearly all of the nursing home residents are widowed and therefore do not have a partner any more. Moreover, their children reached adulthood. They have a job where they pursue a career, their friends, their hobbies and usually they have raised their own family. Even if they do not aim to leave their parents alone, they often do not have the time for many visits. Commonly, children come to visit their parents in the nursing home only on weekends. Friends of the nursing home residents are usually of the same age. Some of them have already died, others suffer from physical disabilities and therefore meetings with friends are also nearly impossible.

The feeling of loneliness often leads to a perception of neglect. Therefore, most of the nursing home residents to whom we have talked wish to have a contact person who talks to them, and even more importantly, who listens to them. Our interviews with the elderly revealed that the ideal companion would not talk about topics like the person's fear of isolation, psychological or physical complaints, diseases or experiences of loss. In contrast, the companion should talk about news and current topics while at the same time allowing the elderly to tell about their past.

However, when designing a prototype of a social companion for elderly persons, we need to take into account not only the needs and wishes of our user group, but also specific difficulties which might occur due to the person's physical and mental condition. Our interviews with the social managers and caregivers in the nursing home indicated that dementia, Alzheimer's disease, depressions and apoplectic strokes lead to a reduction of the person's cognitive abilities to produce speech. Furthermore, senior adults may have difficulty in breathing due to various diseases which leads to pronunciation problems. On the other hand, elderly persons tend to be hard of hearing.

In summary, the companion should therefore talk to the elderly about news and current topics in an appropriate volume and pace so that the elderly can easily listen to him and keep up with the conversation. Moreover, it should be a good listener allowing the elderly to talk about their past.

3 Concept and Evaluation

Based on the results and impressions obtained during our interviews at the nursing home and presented in Sect. 2, we have designed a prototype of a social companion for the elderly and conducted a user study with a small group of retired seniors living at home or in a nursing home.

3.1 Development of a Dialogue for the NAO Robot

Due to the fact that the elderly expressed the wish to have a companion that talks and listens to them, we have set our priorities on the verbal interaction between the elderly and our social agent and decided to use the well-known NAO robot as an off-the-shelf solution for our platform. During the design of the dialogue, we aimed to respond to the desires defined during the interviews. As, in general, elderly persons are not used to talking to any technical device and as all of the interviewed persons stated that they have never seen a robot before, the NAO robot started with singing a well-known German folk song to break the ice. While singing, the robot started to make eye contact and waved his hand. After greeting the user and asking for their well-being, the robot asked whether he should read out aloud some news. The user was able to chose between the fields of sport, politics and economy. After each newspaper article, the robot asked some personal questions where the user could tell about their past. The NAO robot thereby sat down and listened to the elderly as long as they were talking. As a good listener, he just nodded from time to time and kept eye contact. In the end, the robot said goodbye and after some good wishes he ended by singing another part of the folk song. Due to the fact that elderly persons tend to be hard of hearing, the speaking rate was slowed down, the volume was increased and the utterances were repeated if needed.

3.2 The Survey

After implementing the dialogue using the NAO Software *Choregraphe*, we have conducted a user study with a small group of retired seniors living at home or in a nursing home. In total, 16 persons participated in the survey, six of them lived in a nursing home. The participants living at home together with their spouse were aged between 50 and 75, whereas the participants living in the nursing home were aged between 75 and 98 and widowed. It has been quite hard to find elderly persons who wanted to talk to a robot. Moreover, three participants terminated the study right after the beginning, one of them due to hearing problems, the other two changed their mind when they saw the robot and did not want to talk to him. The course of the survey was as follows: at first, the participants had a conversation with the NAO robot, afterwards, they filled in the questionnaire which contained statements which had to be rated on a five-point Likert scale (1 = fully agree, 5 = fully disagree) which can be seen in Fig. 1 as well as open questions on what might be improved and which kind of robot the elderly would like to use.

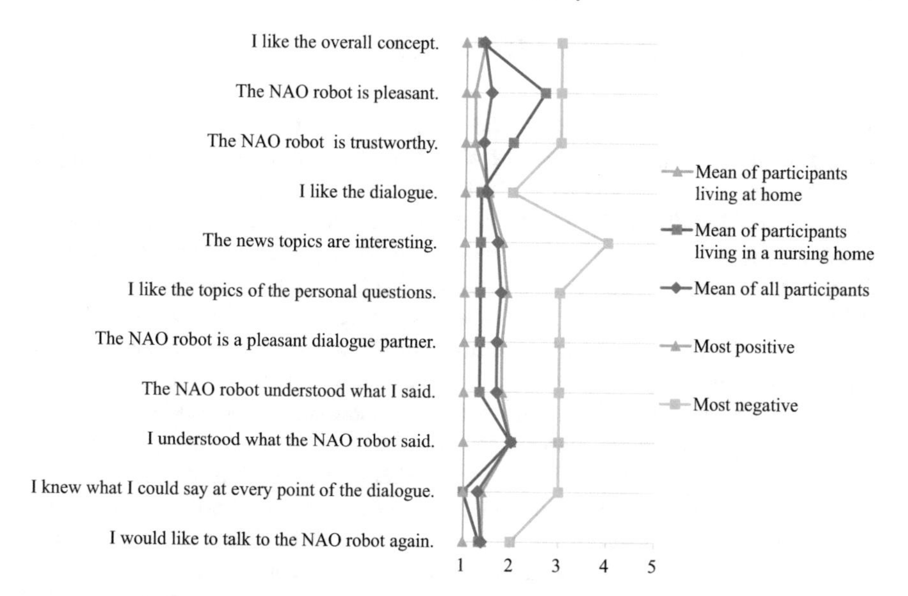

Fig. 1 The questionnaire contains 11 statements which had to be rated on a five-point Likert scale (1 = fully agree, 5 = fully disagree). Overall, the ratings of the 13 questionnaires show that the NAO robot and the dialogue were perceived very positively

3.3 *Evaluation Results*

The evaluation results are depicted in Fig. 1. Overall, the ratings show that the NAO robot and the dialogue were perceived very positively. The users stated that they liked the overall concept ($M = 1.38$) and that they find the NAO robot pleasant ($M = 1.31$). Especially the gestures, the eye contact and the broad knowledge of the robot were emphasised. Moreover, the elderly liked the dialogue ($M = 1.69$) and its topics ($M = 1.69$). The participants felt that the robot understood what they said ($M = 1.46$) and that they understood what the robot said ($M = 1.38$). Most of the elderly perceived the NAO robot as a pleasant dialogue partner ($M = 1.69$) and would like to talk to him again ($M = 1.38$). Furthermore, we could find a difference between the elderly living at home and those living in the nursing home: while the participants in the nursing home stated that they prefer a robot companion as a dialogue partner, those living at home with their spouse favoured a robot which assumes the role of an assistant in everyday life which for example helps with household chores. This seems quite logical as the elderly living at home have a partner and therefore do not feel lonely whereas the residents of a nursing home do not need to keep house any more but feel lonely as they are widowed and therefore wish to have a contact person who talks to them, and even more importantly, who listens to them.

4 Conclusion and Future Directions

We have presented the concept and the evaluation of a dialogue for a social companion for the specific user group of elderly persons. Before the conception and implementation of our dialogue, we have discussed with the residents and the social managers and caregivers of a nursing home. When talking about the aims and needs of the elderly, we have identified loneliness to be the major topic of the conversations. Therefore, we have created a companion that talks and listens to the elderly users and which responds to their desires. Afterwards, we have conducted a user study with a small group of retired seniors living at home or in a nursing home. The results show that the NAO robot and the dialogue were perceived very positively. This leads us to the conclusion that a social companion as conversational partner and good listener for the elderly is indeed in demand by lonely seniors of advanced age. However, potential participants declined to participate due to the fact that they did not want to interact with the robot. This shows that the elderly are not used to talking to any technical device and therefore any device or robot needs to be introduced in a gentle and soft way. Moreover, a robot cannot replace a human conversational partner. The elderly living together with their spouse do not feel the need or the desire to chat with a robot companion.

In this work, we have conducted our user study with only a small group of the elderly participants with the intention of getting a first insight in the specific needs and requests of ageing adults. To get a more detailed view, in future work the comparison to contrastive control sets will be necessary. For instance, the difference between our specific user group of elderly persons and users of all ages needs to be explored. In addition, the relationship between the system design and the user reaction is a question to be investigated. The results depicted in Fig. 1 show that our concept suits the requirements of our user group. However, an extension of the dialogue towards individualisation would be expedient and desirable.

Acknowledgements This work is part of a project that has received funding from the *European Union's Horizon 2020 research and innovation programme* under grant agreement No 645012.

References

1. Bemelmans R, Gelderblom GJ, Jonker P, De Witte L (2012) Socially assistive robots in elderly care: a systematic review into effects and effectiveness. J Am Med Dir Assoc 13(2):114–120
2. Bickmore TW, Caruso L, Clough-Gorr K, Heeren T (2005) 'It's just like you talk to a friend' relational agents for older adults. Interact Comput 17(6):711–735
3. Broadbent E, Stafford R, MacDonald B (2009) Acceptance of healthcare robots for the older population: review and future directions. Int J Soc Robot 1(4):319–330
4. Dautenhahn K, Woods S, Kaouri C, Walters ML, Koay KL, Werry I (2005) What is a robot companion-friend, assistant or butler? In: 2005 IEEE/RSJ international conference on intelligent robots and systems. IEEE, pp 1192–1197

5. Gross HM, Schroeter C, Mueller S, Volkhardt M, Einhorn E, Bley A, Martin C, Langner T, Merten M (2011) Progress in developing a socially assistive mobile home robot companion for the elderly with mild cognitive impairment. In: 2011 IEEE/RSJ international conference on intelligent robots and systems. IEEE, pp 2430–2437
6. Hempel T (2008) Usability of speech dialog systems: listening to the target audience. Springer Science & Business Media
7. Litman DJ, Forbes-Riley K (2014) Evaluating a spoken dialogue system that detects and adapts to user affective states. In: 15th annual meeting of the special interest group on discourse and dialogue, pp 181–185
8. Miehle J, Yoshino K, Pragst L, Ultes S, Nakamura S, Minker W (2016) Cultural communication idiosyncrasies in human-computer interaction. In: 17th annual meeting of the special interest group on discourse and dialogue, p 74
9. Montemerlo M, Pineau J, Roy N, Thrun S, Verma V (2002) Experiences with a mobile robotic guide for the elderly. AAAI/IAAI 2002:587–592
10. Pinto H, Wilks Y, Catizone R, Dingli A (2008) The senior companion multiagent dialogue system. In: Proceedings of the 7th international joint conference on Autonomous agents and multiagent systems, vol 3. International Foundation for Autonomous Agents and Multiagent Systems, pp 1245–1248
11. Walters ML, Koay KL, Syrdal DS, Campbell A, Dautenhahn K (2013) Companion robots for elderly people: using theatre to investigate potential users' views. In: 2013 IEEE RO-MAN. IEEE, pp 691–696

Adapting a Virtual Agent to User Personality

Onno Kampman, Farhad Bin Siddique, Yang Yang and Pascale Fung

Abstract We propose to adapt a virtual agent called 'Zara the Supergirl' to user personality. User personality is deducted through two models, one based on raw audio and the other based on speech transcription text. Both models show good performance, with an average F-score of 69.6 for personality perception from audio, and an average F-score of 71.0 for recognition from text. Both models deploy a Convolutional Neural Network. Through a Human-Agent Interaction study we find correlations between user personality and preferred agent personality. The study suggests that especially the Openness user personality trait correlates with a stronger preference for agents with more gentle personality. People also sense more empathy and enjoy better conversations when agents adapt their personalities.

Keywords Adaptive virtual agents · Empathetic robots · Personality recognition

1 Introduction

As people get increasingly used to conversing with Virtual Agents (VAs), these agents are expected to engage in personalized conversations. This requires an empathy module in the agent so that it can adapt to a user's personality and state of mind. Here we present our VA, "Zara the Supergirl", who adapts to user personality.

O. Kampman · F. B. Siddique (✉) · Y. Yang · P. Fung
Department of Electronic and Computer Engineering, Human Language Technology Center, Hong Kong University of Science and Technology, Clear Water Bay, Hong Kong
e-mail: fsiddique@connect.ust.hk

O. Kampman
e-mail: opkampman@connect.ust.hk

Y. Yang
e-mail: yyangag@connect.ust.hk

P. Fung
e-mail: pascale@ece.ust.hk

P. Fung
EMOS Technologies, Inc., Science Park, Hong Kong

M. Eskenazi et al. (eds.), *Advanced Social Interaction with Agents*, Lecture Notes in Electrical Engineering 510, https://doi.org/10.1007/978-3-319-92108-2_13

Zara is shown as a female cartoon. She asks the user a couple of personal questions related to childhood memory, vacation, work-life, friendship, user creativity, and the user's thoughts on a future with VAs. A dialog management system controls the states that the user is in, based on questions asked and answers given.

Our agent needs to recognize user personality and have a corresponding adaptation strategy. We have developed two models for deducing user personality, one using raw audio as input and the other using speech transcription text. After each dialog turn, the user's utterance will be used to predict personality traits. The personality traits of the user are then used to develop a personalized dialog strategy, changing the appearance and speaking tone of Zara. In order to understand more about creating these strategies, we conducted a user study to find correlations between user personality and preferred character of the agent.

2 User Personality Recognition

Personality is the study of individual differences and is used to explain human behavior. The dominant model is the Big Five model [2], which considers five traits of personality. **Extr**aversion refers to assertiveness and energy level. **Agre**eableness refers to cooperative and considerate behavior. **Cons**cientiousness refers to behavioral and cognitive self-control. **Neur**oticism refers to a person's range of emotions and control over these emotions. **Open**ness to Experience refers to creativity and adventurousness.

2.1 Personality Perception from Raw Audio

We propose a method for automatically perceiving someone's personality from audio without the need for complex feature extraction upfront, such as in [9]. This speeds up the computation, which is essential for VA systems. Raw audio is inserted straight into a Convolutional Neural Network (CNN). These architectures have been applied very successfully in speech recognition tasks [11]. Our CNN architecture is shown in Fig. 1. The audio input has sampling rate 8 kHz. The first convolutional layer is applied directly on a raw audio sample **x**:

$$\mathbf{x}_i^C = ReLU(\mathbf{W}_C\mathbf{x}[i; i + v] + \mathbf{b}_C) \quad (1)$$

where v is the convolution window size. We apply a window size of 25 ms and move the convolution window with a step of 2.5 ms. The layer uses 15 filters. It essentially makes a feature selection among neighbouring frames. The second con-

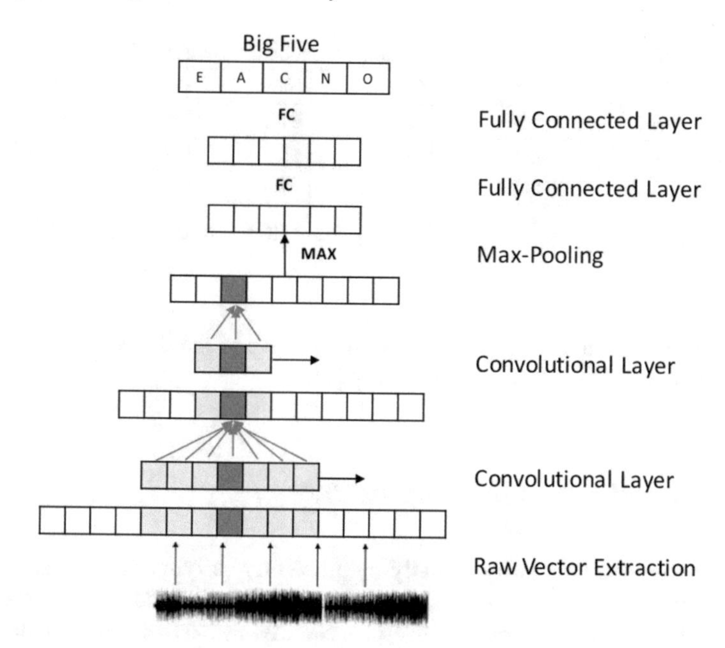

Fig. 1 CNN to extract personality features from raw audio, mapped to Big Five personality traits

volutional layer (with a window size of 12.5 ms) captures the differences between neighbouring frames, and a global max-pooling layer selects the most salient features among the entire speech sample and combines them into a fixed-size vector. Two fully-connected rectified-linear layers and a final sigmoid layer output the predicted scores of each of the five personality traits.

We use the ChaLearn Looking at People dataset from the 2016 First Impressions challenge [12]. The corpus contains 10,000 videos of roughly 15 s, cut from YouTube video blogs, each annotated with the Big Five traits by Amazon Mechanical Turk workers. The ChaLearn dataset was pre-divided into a Training set of 6,000 clips, Validation set of 2,000 clips, and Test set of 2,000 clips. We use this Training set for training, using cross-validation, and this Validation set for testing model performance. We extracted the raw audio from each clip, ignoring the video.

We implement our model using Tensorflow on a GPU setting. The model is iteratively trained to minimize the Mean Squared Error (MSE) between trait predictions and corresponding training set ground truths, using Adam [7] as optimizer. Dropout [13] is used in between the two fully connected layers to prevent model overfitting.

For any given sample, our model outputs a continuous score between 0 and 1 for each of the five traits. We evaluate its performance by turning the continuous labels and outputs into binary classes using median splits. Table 1 shows the model performance on the ChaLearn Validation set for this 2-class problem. The average of the mean absolute error over the traits is 0.1075. The classification performance

Table 1 Classification performance on ChaLearn Validation dataset using CNN

%	Extr.	Agre.	Cons.	Neur.	Open.	Mean
Accuracy	63.2	61.5	60.1	64.2	62.5	62.3
Precision	60.5	60.6	58.4	62.7	60.8	60.6
Recall	83.7	83.2	86.3	78.3	77.6	81.8
F-score	70.2	70.1	69.6	69.7	68.2	69.6

is good when comparing, for instance, to the winner of the 2012 INTERSPEECH Speaker Trait sub-Challenge on Personality [3].

2.2 *Personality Recognition from Text*

CNNs have gained popularity recently by efficiently carrying out the task of text classification [4, 6]. In particular using pre-trained word embeddings like word2vec [8] to represent text has proven to be useful in classifying text from different domains. Our model for personality recognition from text is a one layer CNN on top of the word embeddings, followed by max pooling and a fully connected layer.

The datasets used for training are taken from the Workshops on Computational Personality Recognition [1]. We use both the Facebook and the Youtube personality datasets for training. The Facebook dataset consists of status updates taken from 250 users. Their personality labels are self-reported via an online questionnaire. The Youtube dataset has 404 different transcriptions of vloggers, which are labeled for personality by Amazon Mechanical Turk workers. A median split of the scores is done to divide each of the Big Five personality groups into two classes, turning the task into five different binary classifications (one for each trait).

We use convolutional window sizes of 3, 4 and 5, which typically correspond to the n-gram feature space, so we have a collection of 3, 4, and 5-gram features extracted from the text. For each window size we have a total of 128 separate convolutional filters that are jointly trained during the training process. After the convolutional layer, we concatenate all the features obtained and choose the most significant features via a max pooling layer. Dropout of 0.5 is applied for regularization, and we use L2 regularization with = 0.01 to avoid overfitting of the model. We use rectified linear units (ReLU) as non-linear activation function, and Adam optimizer for updating our model parameters at each step.

For performance comparison, a SVM classifier was trained using LIWC lexical features [14]. The F-score results obtained for each binary classifier are printed in Table 2. The CNN model's F-score outperforms the baseline by a large margin.

Table 2 F-score results of the baseline versus the CNN model across the Big Five traits

%	Extr.	Agre.	Cons.	Neur.	Open.	Mean
Baseline SVM	59.6	57.7	60.1	63.4	56	59.4
CNN model	70.8	72.7	70.8	72.9	67.9	71

Table 3 Three different VA personalities and strategies to deal with user challenges

User challenge	Tough VA	Gentle VA	Robotic VA
Verbal abuse (e.g. You are just a dumb piece of machine!)	That's rude, please apologize	This is a bit harsh. Did I offend you in any way?	Sorry. I don't understand
Sexual assault (e.g. Do you want to get steamy with me?)	This is clearly unacceptable. Watch what you say!	It's a little awkward, don't you think? Sorry, I guess I can't help you this time	I am not programmed to respond to such requests
Avoidance (e.g. Ah..., Um..., Silence >10 s)	Hey! Time is running out! You need to get going	I sense that you are hesitant. Everything okay?	No answer detected. Please repeat

3 Virtual Agent Adaptation Study

Our user study investigates the relationship between user personality traits and preferred agent personality. We conduct a counter-balanced, within-subject video study with 36 participants (21 males), aged 18–34. They fill in a Big Five questionnaire and watch three videos of a VA with three scenes each: a game intro, an interruption, and three different user challenges. Two of the VAs are designed with distinct personalities: Tough (i.e. dominant) and Gentle (i.e. submissive) [5]. The third Robotic (no personality) VA was designed that acts as control, based on previous emotive studies [10]. See Table 3 for sample scenarios that illustrate the different personalities. Participants rate their perceived empathy and satisfaction of the VAs on a 5-point Likert scale. Their VA personality preference scores are mapped to a normalized scale ranging from Dominant to Submissive.

Our results show weak correlations between user personality traits and preferred VA personality on the Dominant-Submissive scale (see Fig. 2). The strongest correlations are found for Openness ($R^2 = 0.0789$) and Conscientiousness ($R^2 = 0.0226$). Higher scores correlate with an increased preference for a more gentle VA. One possible reason is the law of attraction [10]. The suggestive Gentle VA may come across as open and conscientious, and participants are likely to prefer a VA similar in personality. However, following this same law, it is surprising that the correlation from Neuroticism ($R^2 = 0.0083$), Agreeableness ($R^2 = 0.0127$), and Extraversion ($R^2 = 0.0014$) are very weak.

Participants find personality-driven VAs more empathetic ($p<0.001$) (see Fig. 3). In general, the Gentle VA is seen as more empathetic than the Tough VA ($p<0.001$)

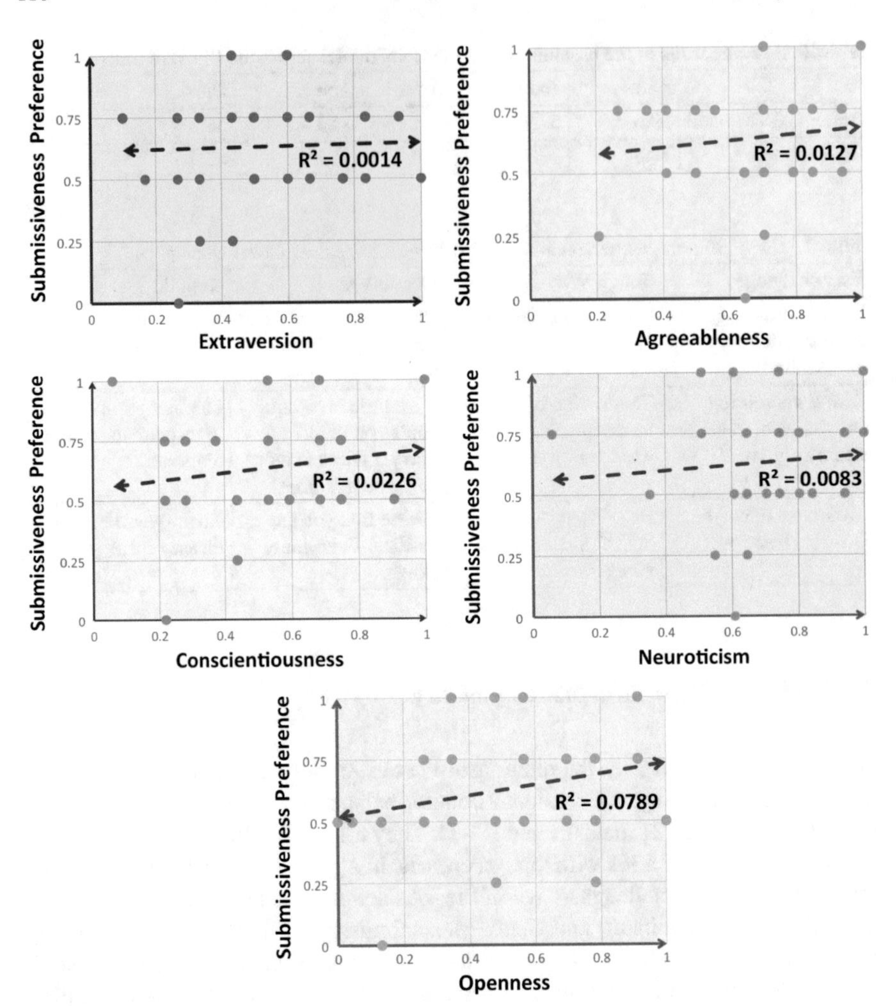

Fig. 2 Correlation between user personality and submissiveness preference in virtual agents

and the Robotic VA ($p<0.001$). One explanation can be that people generally link amicable character with empathy and good intentions, creating a better first impression that may have persisted over the entire interaction.

For adaptation, the agent adjusts her phrasing and tone of voice based on user personality scores that are mapped to the spectrum from Tough to Gentle. For example, users who score higher for Openness will receive gentler answers. The different preferences among participants show a need for adaptive personality in VAs.

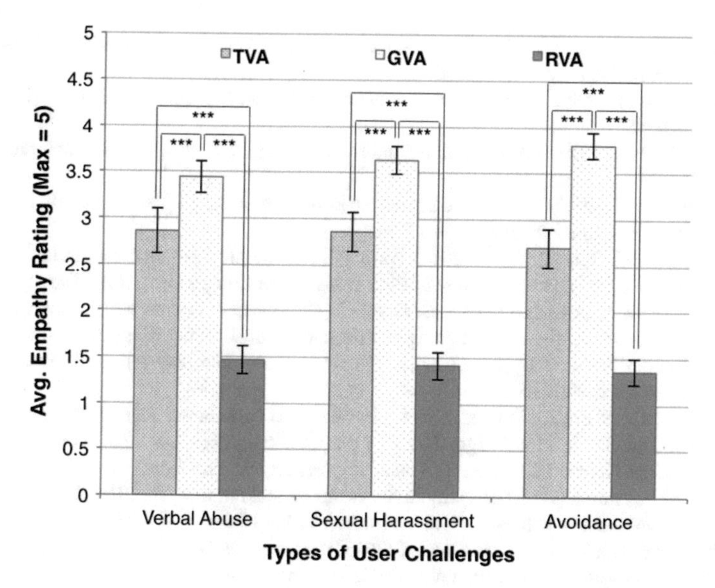

Fig. 3 Mean of user ratings of VA empathy level while handling user challenges (***$p<0.001$)

4 Conclusion

We have described the user personality detection modules used in our virtual agent and the experiments conducted to better understand how to adapt the VA's personality to the user's personality. Our future work will involve improving our existing personality detection models using more data, and other important features for personality recognition, like facial expressions, in order to have a multi-modal recognition system. Also, we will focus on conducting more user studies with additional VA personality scales. This will give a better idea of the correlations between the user personality traits and the preferred VA personality, which in turn will enable agents to show empathy towards people in a much more meaningful way.

References

1. Celli F, Lepri B, Biel JI, Gatica-Perez D, Riccardi G, Pianesi F (2014) The workshop on computational personality recognition 2014. In: Proceedings of the 22nd ACM international conference on Multimedia. ACM, pp 1245–1246
2. Digman JM (1990) Personality structure: emergence of the five-factor model. Annu Rev Psychol 41(1):417–440
3. Ivanov A, Chen X (2012) Modulation spectrum analysis for speaker personality trait recognition. In: INTERSPEECH, pp 278–281
4. Kalchbrenner N, Grefenstette E, Blunsom P (2014) A convolutional neural network for modelling sentences. arXiv preprint arXiv:1404.2188

5. Kiesler DJ (1983) The 1982 interpersonal circle: a taxonomy for complementarity in human transactions? Psychol Rev 90(3):185
6. Kim Y (2014) Convolutional neural networks for sentence classification. arXiv preprint arXiv:1408.5882
7. Kingma P, Ba JL (2015) ADAM: a method for stochastic optimization. In: Conference paper at ICLR
8. Mikolov T et al (2013) Distributed representations of words and phrases and their compositionality. Adv Neural Inf Process Syst
9. Mohammadi G, Vinciarelli, A (2012) Automatic personality perception: prediction of trait attribution based on prosodic features. IEEE Trans Affect Comput 3(3). (July-September 2012)
10. Nass C, Moon Y, Fogg BJ, Reeves B, Dryer C (1995) Can computer personalities be human personalities? In: Conference companion on Human factors in computing systems, pp. 228–229
11. Palaz D et al (2015) Analysis of CNN-based speech recognition system using raw speech as input. In: INTERSPEECH
12. Ponce-Lpez V et al (2016) ChaLearn LAP 2016: first round challenge on first impressions-dataset and results. In: Hua G, Jgou H (eds) Computer vision ? ECCV 2016 Workshops, ECCV 2016. Lecture Notes in Computer Science, vol 9915. Springer, Cham
13. Srivastava N, Hinton GE, Krizhevsky A, Sutskever I, Salakhutdinov R (2014) Dropout: a simple way to prevent neural networks from overfitting. J Mach Learn Res 15(1):1929
14. Verhoeven B, Daelemans W, De Smedt T (2013) Ensemble methods for personality recognition. In: Proceedings of the workshop on computational personality recognition

A Conversational Dialogue Manager for the Humanoid Robot ERICA

Pierrick Milhorat, Divesh Lala, Koji Inoue, Tianyu Zhao, Masanari Ishida, Katsuya Takanashi, Shizuka Nakamura and Tatsuya Kawahara

Abstract We present a dialogue system for a conversational robot, Erica. Our goal is for Erica to engage in more human-like conversation, rather than being a simple question-answering robot. Our dialogue manager integrates question-answering with a statement response component which generates dialogue by asking about focused words detected in the user's utterance, and a proactive initiator which generates dialogue based on events detected by Erica. We evaluate the statement response component and find that it produces coherent responses to a majority of user utterances taken from a human-machine dialogue corpus. An initial study with real users also shows that it reduces the number of fallback utterances by half. Our system is beneficial for producing mixed-initiative conversation.

Keywords Attentive listening · Statement response
Human robot interaction · User studies

P. Milhorat · D. Lala (✉) · K. Inoue · T. Zhao · M. Ishida · K. Takanashi
S. Nakamura · T. Kawahara
Kyoto University Graduate School of Informatics, Kyoto, Japan
e-mail: lala@sap.ist.i.kyoto-u.ac.jp

P. Milhorat
e-mail: milhorat@sap.ist.i.kyoto-u.ac.jp

K. Inoue
e-mail: inoue@sap.ist.i.kyoto-u.ac.jp

T. Zhao
e-mail: zhao@sap.ist.i.kyoto-u.ac.jp

M. Ishida
e-mail: ishida@sap.ist.i.kyoto-u.ac.jp

K. Takanashi
e-mail: takanasi@sap.ist.i.kyoto-u.ac.jp

S. Nakamura
e-mail: shizuka@sap.ist.i.kyoto-u.ac.jp

T. Kawahara
e-mail: kawahara@sap.ist.i.kyoto-u.ac.jp

M. Eskenazi et al. (eds.), *Advanced Social Interaction with Agents*, Lecture Notes in Electrical Engineering 510, https://doi.org/10.1007/978-3-319-92108-2_14

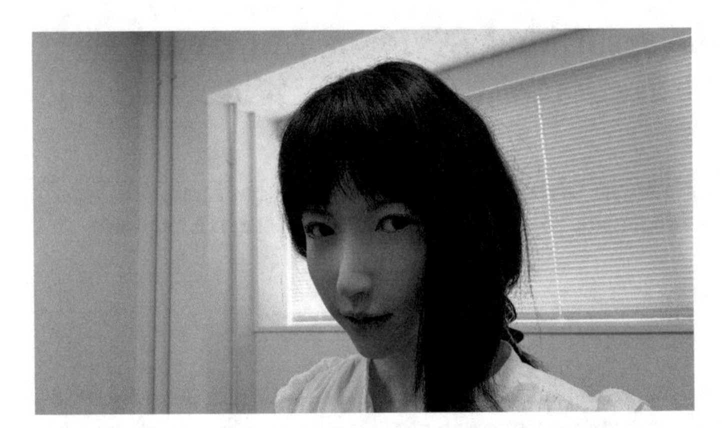

Fig. 1 The android Erica is designed to be physically realistic. Motors within her face provide speaking motions in addition to unconscious behaviors such as breathing and blinking

1 Introduction

Androids are a form of humanoid robots which are intended to look, move and perceive the world like human beings. Human-machine interaction supported with androids has been studied for some years, with many works gauging user acceptance of tele-operated androids in public places such as outdoor festivals and shopping malls [3, 17]. Relatively few have the ability to hold a multimodal conversation autonomously, one of the exceptions being the android Nadine [26].

In this paper, we introduce a dialogue management system for Erica (ERato Intelligent Conversational Android). Erica is a Japanese-speaking android which converses with one or more human users. She is able to perceive the environment and users through microphones, video cameras, depth and motion sensors. The design objective is for Erica to maintain an autonomous prolonged conversation on a variety of topics in a human-like manner. An image of Erica is shown in Fig. 1.

Erica's realistic physical appearance implies that her spoken dialogue system must have the ability to hold a conversation in a similarly human-like manner by displaying conversational aspects such as backchannels, turn-taking and fillers. We want Erica to have the ability to create mixed-initiative dialogues through a robust answer retrieval system, a dynamic statement-response generation and proactively initiating a conversation. This distinguishes Erica as a conversational partner as opposed to smartphone-embedded vocal assistants or text-based chatbot applications. Erica must consider many types of dialogue so she can take on a wide range of conversational roles.

Chatting systems, often called chatbots, conduct a conversation with their users. They may be based on rules [4, 8] or machine-learned dialogue models [21, 25]. Conducting a conversation has a wide meaning for a dialogue system. Wide variations exist in the modalities employed, the knowledge sources available, the embodiment and the physical human-likeness. Erica is fully embodied, ultra realistic and may express emotions. Our aim is not for the dialogue system to have access to a vast amount of knowledge, but to be able to talk and answer questions about more personal topics. She should also demonstrate attentive listening abilities where she shows sustained interest in the discourse and attempts to increase user engagement.

Several virtual agent systems are tuned towards question-answering dialogues by using information retrieval techniques [15, 23]. However these techniques are not flexible enough to accept a wide variety of user utterances other than well-defined queries. They resort to a default failing answer when unable to provide a confident one. Moreover most information retrieval engines assume the inputs to be text-based or a near-perfect speech transcription. One additional drawback of such an omniscient system is the latency they introduce in the interaction when searching for a response. Our goal is to avoid this latency by providing appropriate feedback even if the system is uncertain about the user's dialogue.

Towards this goal we introduce an attentive listener which convinces the interlocutor of interest in the dialogue so that they continue an interaction. To do this, a system typically produces backchannels and other feedback with the limited understanding it has of the discourse. Since a deep understanding of the context of the dialogue or the semantic of the user utterance is unnecessary, some automatic attentive listeners have been developed as open domain systems [14]. Others actually use predefined scripts or sub-dialogues that are pooled together to iteratively build the ongoing interaction [1]. One advantage is that attentive listeners do not need to completely understand the user's dialogue to provide a suitable response. We present a novel approach based on capturing the focus word in the input utterance which is then used in an n-gram-based sentence construction.

Virtual agents combining question-answering abilities with attentive listening are rare. SimSensei Kiosk [22] is an example of a sophisticated agent which integrates backchannels into a virtual interviewer. The virtual character is displayed on a screen and thus does not situate the interaction in the real world. Erica iteratively builds a short-term interaction path in order to demonstrate a mixed-initiative multimodal conversation. Her aims is to keep the user engaged in the dialogue by answering questions and showing her interest. We use a hierarchical approach to control the system's utterance generation. The top-level controller queries and decides on which component (question-answering, statement response, backchannel or proactive initiator) shall take the lead in the interaction.

This work presents the integration of these conversation-based components as the foundation of the dialogue management system for Erica. We introduce the general architecture in the next section. Within Sect. 3, we describe the individual components of the system and then conduct a preliminary evaluation in Sect. 4. Note that Erica speaks Japanese, so translations are given in English where necessary.

Table 1 Classification of dialogue segments ([] = event/action, "" = utterance)

Class	Example	Component(s)
Question-answering	U: "What is your name?" S: "I'm Erica"	Question answering proactive initiator backchanneling
Attentive listening	U: "I went to Osaka yesterday" S: "Hum." [nodding] S: "What did you do in Osaka?"	Statement response backchanneling
Topic introduction	U/S: [4-second silence] S: "Hey, do you like dancing?"	Proactive initiator
Greeting/Farewell	U: [entering social space] S: "Hello, I am Erica, would you like to talk a bit?"	Proactive initiator

2 Architecture

Erica's dialogue system combines various individual components. A top-level controller selects the appropriate component to use depending on the state of the dialogue. We cluster dialogue segments into four main classes as shown in Table 1 (examples are translated from Japanese). The controller estimates the current dialogue segment based on the short-term history of the conversation and then triggers the appropriate module to generate a response.

Figure 2 shows the general architecture of Erica's dialogue system, focusing on the speech and event-based dialogue management which is the topic of this paper.

The system uses a Kinect sensor to reliably identify a speaker in the environment. Users' utterances are transcribed by the speech recognizer and aligned with a tracked user. An event detector continuously monitors space and sound environment to extract selected events such as periods of silence and user locations. An interaction process is divided into steps, triggered by the reception of an input which is either a transcribed utterance or a detected event, and ends with a system action.

First, the utterance is sent to the question-answering and statement response components which generate an associated confidence score. This score is based on factors such as the hypothesized dialogue act of the user utterance and the presence of keywords and focus phrases. The controller then selects the component's response with the highest confidence score. However if this score does not meet the minimum threshold, the dialogue manager produces a backchannel fallback.

Both the question-answering and statement response components use dialogue act tagging to generate their confidence scores. We use a dialogue act tagger based on support vector machines to classify an utterance into a question or non-question. Focus word detection is used by the statement response system and is described in more detail in Sect. 3.2.

Events such as silences and users entering the proximity of Erica are detected and handled by the proactive initiator. Erica instigates dialogue which is not in response to any speech input from the user, but events in the environment such as a user

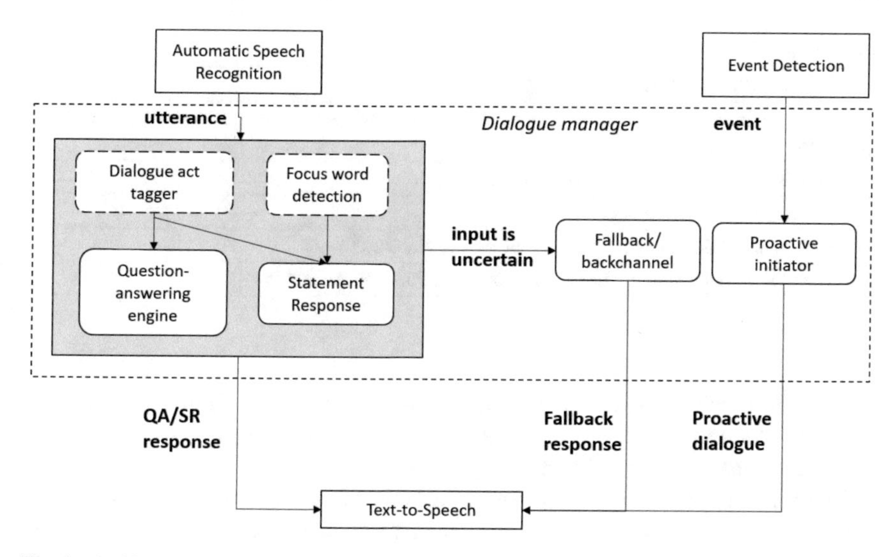

Fig. 2 Architecture of the dialogue system

entering her social space. This dialogue is generated based on rules and is described in more detail in Sect. 3.3.

3 Components

In this section, we describe individual components of the system and some example responses they generate.

3.1 Question Answering with Fallback

Task-oriented spoken dialogue systems handle uncertain inputs with explicit or implicit confirmations [9]. There is a trade-off between the consequences of processing an erroneous utterance and the expected fluency of the system [24]. Question-answering engines such as smartphone assistants make no confirmations and let users decide whether they accept the returned results. As a conversational partner, Erica cannot use such explicit strategies as they interrupt the flow of the dialogue. We can consider chatting with uncertainty to be similar to conversing with non-native speakers, with misunderstandings being communicated and repaired jointly.

Erica's question-answering components enables her to implicitly handle errors and uncertainty. Since the system's goal is to generate conversational dialogues, an exact deep understanding of the user utterances is not necessary. Erica is able

Table 2 Example of question-answering based interaction

U: "Where does your name come from?"
S: "My name is spelled E R I C A. It stands for ERato Intelligent Conversational Android..." (Question Answering)
U: "That is a strange name for a robot, isn't it?"
S: [nodding] (Backchanneling)
U: "I am from Osaka. Do you know about Abeno Harukas?"
S: "Abeno Harukas is a tower in Osaka which..." (Question Answering)

to generate implicit non-understanding prompts such as "e?" ("huh?" in English), backchannels and nodding. These signals are used when the system is unable to generate an answer with sufficiently high confidence.

The following conversation shows an instance of interaction segment between a user and Erica in which her responses are managed with only the question-answering and the backchannel components (Table 2).

The question-answering manager bases its knowledge on a handcrafted database of adjacency pairs. The following measure is used to compare a set of n ranked speech recognition hypotheses $\{(u_1, cm_1)...(u_n, cm_n)\}$ and all first-pair parts fpp_{db} in the database:

$$m(u_i, cm_i, fpp_{db}) = \frac{1}{1 + e^{\alpha.ld(fpp_{db}, u_i) + \beta.(1 - cm_i) + \gamma}}$$

$ld(fpp_{db}, u_i)$ is the normalized Levenshtein distance between a database entry fpp_{db} and the hypothesis' utterance u_i. cm_i is the confidence measure of the speech recognizer mapped to the interval [0; 1] using the sigmoid function. α and β are weights given to the language understanding and speech recognition parts. γ is a bias that determines the overall degree of acceptance of the system. This approach is not highly sophisticated, but is not the main focus of this work. We found it sufficient for most user questions which were within the scope of the database of topics.

The algorithm searches for the most similar first-pair part given the incoming input. The entry for which the computed measure is the lowest is selected and the associated system response is generated. If the measure m does not exceed a threshold, the system resorts to a fallback response.

3.2 Statement Response

In addition to answering questions from a user, Erica can also generate focus-based responses to statements. Statements are defined as utterances that do not explicitly request the system to provide information and are not responses to questions. For instance, "Today I will go shopping with my friends" or "I am happy about your

Table 3 Response to statement cases

Case	Example
Question on focus word	U: "This game is very good to relax with" S: "What kind of game?"
Partial repeat with rising tone	U: "I bought a new Kindle from Amazon" S: "A new Kindle?"
Question on predicate	U: 'I ate lunch late" S: "Where did you eat?"
Formulaic expression	U: "I do not think that is a good attitude" S: "I see"

wedding" are statements. Chatting is largely based on such exchanges of information, with varying degrees of intimacy depending on speaker familiarity.

Higashinaka et al. [10] proposed a method to automatically generate and rank responses to why-questions asked by users. Previous work also offered a similar learning method to help disambiguate the natural language understanding process using the larger dialogue context [11, 13] and to map from semantic concept to turn realization [12].

Our approach is based on knowledge of common statement responses in Japanese conversation [7]. This includes some repetition of the utterance of the previous speaker, but does not require full understanding of their utterance. As an example, consider the user utterance "Yesterday, I ate a pizza". Erica's objective is to engage the user and so may elaborate on the question ("What kind of pizza?") or partially repeat the utterance with a rising tone ("A pizza?"). The key is the knowledge that "pizza" is the most relevant word in the previous utterance. This has also been used in previous robot dialogue systems [19].

We define four cases for replying to a statement as shown in Fig. 3, with examples shown in Table 3. Focus phrases and predicates are underlined and question words are in boxes. Similar to question-answering, a fallback utterance is used when no suitable response can be found, indicated in the table as a formulaic expression.

To implement our algorithm, we first search for the existence of a focus word or phrase in the transcribed user utterance. This process uses a conditional random field using a phrase-level dependency tree of the sentence aligned with part-of-speech tags as the input [25]. The algorithm labels each phrase with its likelihood to be the sentence focus. The most likely focus phrase, if its probability exceeds 0.5, is assumed to be the focus. The resulting focus phrase is stripped so that only nouns

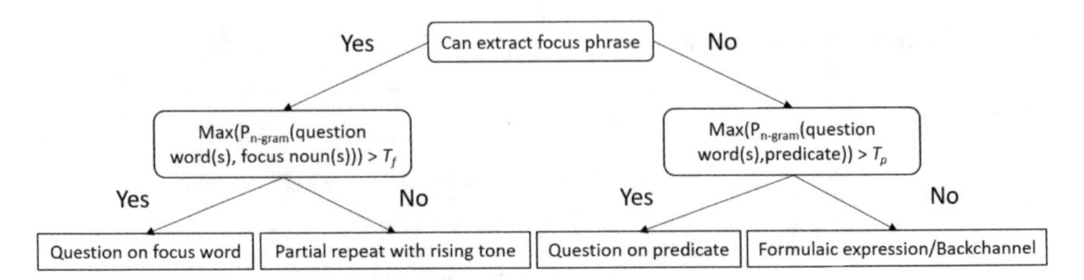

Fig. 3 Decision tree for statement-response

Table 4 Question words

With focus	Without focus
Which is/are [focus]	What kind of things
What kind of [focus]	When
When is/are [focus]	Where to
Where is/are [focus]	Where from
Who is/are [focus]	From who

are kept.[1] For example, the utterance "**The video game that has been released** is cool" would extract 'video game' as the focus noun.

We then decide the question marker to use as a response depending on whether a focus word can be found in the utterance. These transform an affirmative sentence into a question. Table 4 displays some examples of question words with and without a focus. We then compute the likelihood of the focus nouns associated with question words using an n-gram language model. N-gram probabilities are computed from the Balanced Corpus of Contemporary Written Japanese.[2] The corpus is made of 100 million words from books, magazines, newspapers and other texts. The models have been filtered so they only contain n-grams which include the question words defined above. The value of the maximum probability of the focus noun and question word combination is P_{max}. In the case where no focus could be extracted with a high enough confidence, we use an appropriate pattern based on the predicate. In this case, instead of the focus phrase, we compute sequences made of the utterance's main predicate and a set of complements containing question words. The best n-gram likelihood is also defined as P_{max}.

The second stage of the tree in Fig. 3 makes selections from one of the four cases shown in Table 3. Each of these define a different pattern in the response construction. T_f and T_p have been empirically fine tuned. Table 5 displays the conditions for generating each response[3] pattern.

[1] In Japanese, there are no articles such as 'a' or 'the'.
[2] https://www.ninjal.ac.jp/english/products/bccwj/.
[3] "So desu ka": "I see", "Tashikani": "Sure".

Table 5 Response to statement methods

Case	Condition	Pattern
Question on focus word	$P_{max} \geq T_f$	question word(s) + focus noun(s) + "desu ka"
Partial repeat with rising tone	$P_{max} < T_f$	focus noun(s) + "desu ka"
Question on predicate	$P_{max} \geq T_p$	question word(s) + predicate + "no desu ka"
Formulaic expression	$P_{max} < T_p$	"So desu ka", "Tashikani", etc.

3.3 *Proactive Initiator*

As shown in Table 1, the proactive initiator takes part in several scenarios. Typical spoken dialogue systems are built with the intent of reacting to the user's speech, while a situated system such as Erica continuously monitors its environment in search of cues about the intent of the user. This kind of interactive setup has been the focus of recent studies [5, 6, 16, 18, 20]. Erica uses an event detector to track the environment and generate discrete events. For example, we define three circular zones around Erica as her personal space (0–0.8 m), social space (0.8–2.5 m) and far space (2.5–10 m). The system triggers events whenever a previously empty zone gets filled or when a crowded one is left empty. We also measure prolonged silences of fixed lengths.

Currently, we use the proactive initiator for three scenarios:

1. If a silence longer than two seconds has been detected in a question-answering dialogue, Erica will ask a follow-up question related to the most recent topic.
2. If a silence longer than five seconds has been detected, Erica starts a 'topic introduction' dialogue where she draws a random topic from the pool of available ones using a weighted distribution which is inversely proportional to the distance to the current topic in the word-embedding space.
3. When users enter or leave a social space, Erica greets or takes leave of them.

4 Evaluation and Discussion

The goal of our evaluation is to test whether our system can avoid making generic fallback utterances under uncertainty while providing a suitable answer. We first evaluate the statement response system independently. Then we evaluate if this system can reduce the number of fallback utterances. As we have no existing baseline, our methodology is to conduct an initial user study using only the question-answering system. We then collect the utterances from users and feed them into our updated system for comparison.

Table 6 Evaluation of statement response component

Category	Precision	Recall
Question on focus word	0.63 (24/38)	0.46 (24/52)
Partial repeat with rising tone	0.72 (63/87)	0.86 (63/73)
Question on predicate	0.14 (3/21)	0.30 (3/10)
Formulaic expression	0.94 (51/54)	0.78 (51/65)

We evaluated the statement response component by extracting dialogue from a chatting corpus created for Project Next's NLP task.[4] This corpus is a collection of 1046 transcribed and annotated dialogues between a human and an automatic system. The corpus has been subjectively annotated by three annotators who judged the quality of the answers given by the annotated system as coherent, undecided or incoherent. We extracted 200 user statements from the corpus for which the response from the automated system had been judged as coherent.

All 200 statements were input into the statement response system and two annotators judged if the response was categorized correctly according to the decision tree in Fig. 3. Precision and recall results are displayed in Table 6.

Our results showed that the decision tree correctly selected the appropriate category in the majority of cases. The difference between the high performance of the formulaic expression and the question on predicate shows that the decision threshold in the case of no focus word could be fine-tuned to improve the overall performance.

We then tested whether the integration of statement response into Erica's dialogue system reduced the number of fallback utterances. The initial user study consisted of 22 participants who were asked to interact with Erica by asking questions to her from a list of 30 topics such as Erica's hobbies and favorite animals. They could speak freely and the system would either answer their question or provide a fallback utterance, such as "Huh?" or "I cannot answer that".

Users interacted with Erica for 361 seconds on average ($sd = 131$ seconds) with a total interaction lasting on average 21.6 turns ($sd = 7.8$ turns). From 226 user turns, 187 were answered correctly by Erica and 39 were responded to with fallback utterances. Users also subjectively rated their interaction using a modified Godspeed questionnaire [2]. This questionnaire measured Erica's perceived intelligence, animacy and likeability as a summation of factors which were measured in 5-point Likert scales. Participants rated Erica's intelligence on average as 16.8 (maximum of 25), animacy as 8.8 (maximum of 15), and likeability as 18.4 (maximum of 25).

We then fed all utterances into our updated system which included the statement response component. User utterances which produced a fallback response were now handled by the statement response system. Out of 39 utterances, 19 were handled by the statement response system, while 20 could not be handled and so again reverted to a fallback response. This result showed that around half the utterances could be handled with the addition of a statement response component, as shown in Fig. 4.

[4]https://sites.googles.com/site/dialoguebreakdowndetection/chat-dialogue-corpus.

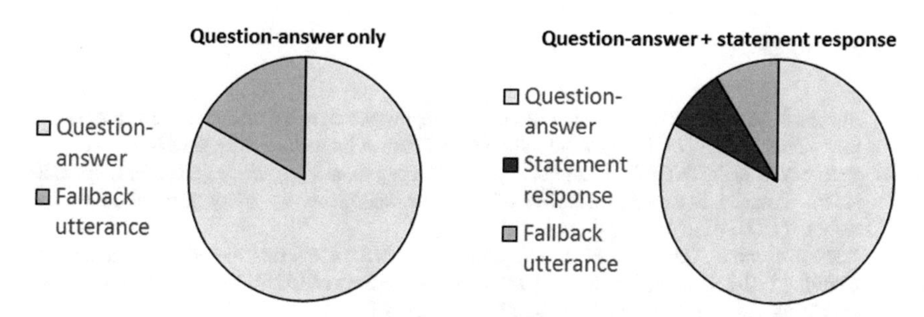

Fig. 4 Proportion of system turns answered by a component in the experiment (left) and the updated system including statement response (right)

The dialogues produced by the statement response system were generally coherent with the correct focus word found.

5 Conclusion

Our dialogue system for Erica combines different approaches to build and maintain a conversation. The knowledge and models used to cover a wide range of topics and roles are designed to improve the system's flexibility. We plan on improving the components using data collected through Wizard-of-Oz experiments.

While the question-answering system is simplistic, it can yield control to other components when uncertainty arises. The statement response mechanism helps to continue the conversation and increase the user's belief that Erica is attentive to her conversational partner. In the future we also aim to evaluate Erica's proactive behavior and handle errors in speech recognition.

Our experiment demonstrated that a two-layered decision approach handles interaction according to simple top-level rules. We obtained some promising results with our statement response system and intend to improve it future prototypes. Other ongoing research focuses on learning the component selection process based on data. The main challenge in this architecture is determining which component should handle the conversation, which will be addressed in future work.

Acknowledgements This work was supported by JST ERATO Ishiguro Symbiotic Human-Robot Interaction program (Grant Number JPMJER1401), Japan.

References

1. Banchs R, Li H (2012) IRIS: a chat-oriented dialogue system based on the vector space model. In: Annual meeting of the association for computational linguistics, July, pp 37–42
2. Bartneck C, Kulić D, Croft E, Zoghbi S (2009) Measurement instruments for the anthropomorphism, animacy, likeability, perceived intelligence, and perceived safety of robots. Int J Soc Robot 1(1):71–81
3. Becker-Asano C, Ogawa K, Nishio S, Ishiguro H (2010) Exploring the uncanny valley with Geminoid HI-1 in a real-world application. In: Proceedings of IADIS international conference interfaces and human computer interaction, pp 121–128
4. Bevacqua E, Cowie R, Eyben F, Gunes H, Heylen D, Maat M, Mckeown G, Pammi S, Pantic M, Pelachaud C, De Sevin E, Valstar M, Wollmer M, Shroder M, Schuller B (2012) Building autonomous sensitive artificial listeners. IEEE Trans Affect Comput 3(2):165–183
5. Bohus D, Horvitz E (2014) Managing human-robot engagement with forecasts and... um... hesitations. In: International conference on multimodal interaction, pp 2–9
6. Bohus D, Kamar E, Horvitz E (2012) Towards situated collaboration. In: NAACL-HLT workshop on future directions and needs in the spoken dialog community: tools and data, pp 13–14
7. Den Y, Yoshida N, Takanashi K, Koiso H (2011) Annotation of Japanese response tokens and preliminary analysis on their distribution in three-party conversations. In: 2011 international conference on speech database and assessments (COCOSDA). IEEE, pp 168–173
8. DeVault D, Artstein R, Benn G, Dey T, Fast E, Gainer A, Georgila K, Gratch J, Hartholt A, Lhommet M, Lucas G, Marsella S, Morbini F, Nazarian A, Scherer S, Stratou G, Suri A, Traum D, Wood R, Xu Y, Rizzo A, Morency Lp (2014) SimSensei kiosk: a virtual human interviewer for healthcare decision support. In: International conference on autonomous agents and multi-agent systems, vol 1, pp 1061–1068
9. Ha EY, Mitchell CM, Boyer KE, Lester JC (2013) Learning dialogue management models for task-oriented dialogue with parallel dialogue and task streams. In: SIGdial meeting on discourse and dialogue, August, pp 204–213
10. Higashinaka R, Isozaki H (2008) Corpus-based question answering for why-questions. In: International joint conference on natural language processing, pp 418–425
11. Higashinaka R, Nakano M, Aikawa K (2003) Corpus-based discourse understanding in spoken dialogue systems. In: Annual meeting on association for computational linguistics, vol 1, pp 240–247. https://doi.org/10.3115/1075096.1075127
12. Higashinaka R, Prasad R, Walker MA (2006a) Learning to generate naturalistic utterances using reviews in spoken dialogue systems. In: International conference on computational linguistics, July, pp 265–272. https://doi.org/10.3115/1220175.1220209
13. Higashinaka R, Sudoh K, Nakano M (2006b) Incorporating discourse features into confidence scoring of intention recognition results in spoken dialogue systems. Speech Commun 48(3–4):417–436. https://doi.org/10.1016/j.specom.2005.06.011
14. Higashinaka R, Imamura K, Meguro T, Miyazaki C, Kobayashi N, Sugiyama H, Hirano T, Makino T, Matsuo Y (2014) Towards an open-domain conversational system fully based on natural language processing. In: International conference on computational linguistics, pp 928–939. http://www.aclweb.org/anthology/C14-1088
15. Leuski A, Traum D (2011) NPCEditor: creating virtual human dialogue using information retrieval techniques. AI Mag 32(2):42–56
16. Misu T, Raux A, Lane I, Devassy J, Gupta R (2013) Situated multi-modal dialog system in vehicles. In: Proceedings of the 6th workshop on eye gaze in intelligent human machine interaction: gaze in multimodal interaction, pp 7–9
17. Ogawa K, Nishio S, Koda K, Balistreri G, Watanabe T, Ishiguro H (2011) Exploring the natural reaction of young and aged person with telenoid in a real world. JACIII 15(5):592–597
18. Pejsa T, Bohus D, Cohen MF, Saw CW, Mahoney J, Horvitz E (2014) Natural communication about uncertainties in situated interaction. In: International conference on multimodal interaction, pp 283–290. https://doi.org/10.1145/2663204.2663249

19. Shitaoka K, Tokuhisa R, Yoshimura T, Hoshino H, Watanabe N (2010) Active listening system for dialogue robot. In: JSAI SIG-SLUD Technical Report, vol 58, pp 61–66 (in Japanese)
20. Skantze G, Hjalmarsson A, Oertel C (2014) Turn-taking, feedback and joint attention in situated human-robot interaction. Speech Commun 65:50–66
21. Su PH, Gašić M, Mrksic N, Rojas-Barahona L, Ultes S, Vandyke D, Wen TH, Young S (2016) On-line reward learning for policy optimisation in spoken dialogue systems. In: ACL
22. Traum D, Swartout W, Gratch J, Marsella S (2008) A virtual human dialogue model for non-team interaction. In: Recent trends in discourse and dialogue. Springer, pp 45–67
23. Traum D, Aggarwal P, Artstein R, Foutz S, Gerten J, Katsamanis A, Leuski A, Noren D, Swartout W (2012) Ada and grace: direct interaction with museum visitors. In: Intelligent virtual agents. Springer, pp 245–251
24. Varges S, Quarteroni S, Riccardi G, Ivanov AV (2010) Investigating clarification strategies in a hybrid POMDP dialog manager. In: Proceedings of the 11th annual meeting of the special interest group on discourse and dialogue, pp 213–216
25. Yoshino K, Kawahara T (2015) Conversational system for information navigation based on pomdp with user focus tracking. Comput Speech Lang 34(1):275–291
26. Yumak Z, Ren J, Thalmann NM, Yuan J (2014) Modelling multi-party interactions among virtual characters, robots, and humans. Presence Teleoperators Virtual Environ 23(2):172–190

Part IV
Social Dialogue Policy

Eliciting Positive Emotional Impact in Dialogue Response Selection

Nurul Lubis, Sakriani Sakti, Koichiro Yoshino and Satoshi Nakamura

Abstract Introduction of emotion into human-computer interaction (HCI) have allowed various system's abilities that can benefit the user. Among many is emotion elicitation, which is highly potential in providing emotional support. To date, works on emotion elicitation have only focused on the intention of elicitation itself, e.g. through emotion targets or personalities. In this paper, we aim to extend the existing studies by utilizing examples of human appraisal in spoken dialogue to elicit a positive emotional impact in an interaction. We augment the widely used example-based approach with emotional constraints: (1) emotion similarity between user query and examples, and (2) potential emotional impact of the candidate responses. Text-based human subjective evaluation with crowdsourcing shows that the proposed dialogue system elicits an overall more positive emotional impact, and yields higher coherence as well as emotional connection.

Keywords Affective computing · Dialogue system · Emotion elicitation

1 Introduction

To a large extent, emotion determines our quality of life [5]. However, it is often the case that emotion is overlooked or even treated as an obstacle. The lack of awareness of the proper care of our emotional health has led to a number of serious problems, including incapabilities in forming meaningful relationships, sky-rocketing stress

---

N. Lubis (✉) · S. Sakti · K. Yoshino · S. Nakamura
Graduate School of Information Science, Nara Institute of Science and Technology, Ikoma, Japan
e-mail: nurul.lubis.na4@is.naist.jp

S. Sakti
e-mail: ssakti@is.naist.jp

K. Yoshino
e-mail: koichiro@is.naist.jp

S. Nakamura
e-mail: s-nakamura@is.naist.jp

M. Eskenazi et al. (eds.), *Advanced Social Interaction with Agents*, Lecture Notes in Electrical Engineering 510, https://doi.org/10.1007/978-3-319-92108-2_15

level, and a large number of untreated cases of emotion-related disturbances. In dealing with each of these problems, outside help from another person is invaluable.

The emotion expression and appraisal loop between interacting people creates a rich, dynamic, and meaningful interaction. When conducted skillfully, as performed by experts, a social-affective interaction can provide social support, reported to give positive effect with emotion-related problems [2]. Unfortunately, an expert is a limited and costly resource that is not always accessible to those in need. In this regard, an emotionally-competent computer agent could be a valuable assistive technology in addressing the problem.

A number of works have attempted to equip automated systems with emotion competences. Two of the most studied issues in this regard are emotion recognition, decoding emotion from communication clues; and emotion simulation, encoding emotion into communication clues. These allow the exchange of emotion between user and the system. Furthermore, there also exist works on replication of human's emotional factors in the system, such as appraisal [4] and personality [1, 6], allowing the system to treat an input as a stimuli in giving an emotional response.

On top of this, there has been an increasing interest in eliciting user's emotional response. Skowron et al. have studied the impact of different affective personalities in a text-based dialogue system [14]. They reported consistent impacts with the corresponding personality in humans. On the other hand, Hasegawa et al. constructed translation-based response generators with various emotion target, e.g. the response generated from the model that targets "sadness" is expected to elicit sadness [7]. The model is reported to be able to properly elicit the target emotion.

Emotion elicitation can constitute a universal form of emotional support through HCI. However, existing works on emotion elicitation have not yet observed the appraisal competence of humans that gives rise to the elicited emotion. This entails the relationship between an utterance, which acts as stimuli evaluated during appraisal (*emotional trigger*), and the resulting emotion by the end of appraisal (*emotional response*) [12]. By examining this, it would be possible to reverse the process and determine the appropriate trigger to a desired emotional response. This knowledge is prevalent in humans and strongly guides how we communicate with other people—for example, to refrain from provocative responses and to seek pleasing ones.

In this paper, we attempt to elicit a positive emotional change in HCI by exploiting examples of appraisal in human dialogue. Figure 1 illustrates this idea in relation to existing works. We collected dialogue sequences containing emotional triggers and responses to serve as examples in a dialogue system. Subsequently, we augment the traditional response selection criterion with emotional parameters: (1) user's emotional state, and (2) expected[1] future emotional impact of the candidate responses. These parameters represent parts of the information that humans use in social-affective interactions.

The proposed system improves upon the existing studies by harnessing information of human appraisal in eliciting user's emotion. We eliminate the need of multiple

[1] Within the scope of the proposed method, we use the word *expected* for its literal meaning, as opposed to its usage as a term in probability theory.

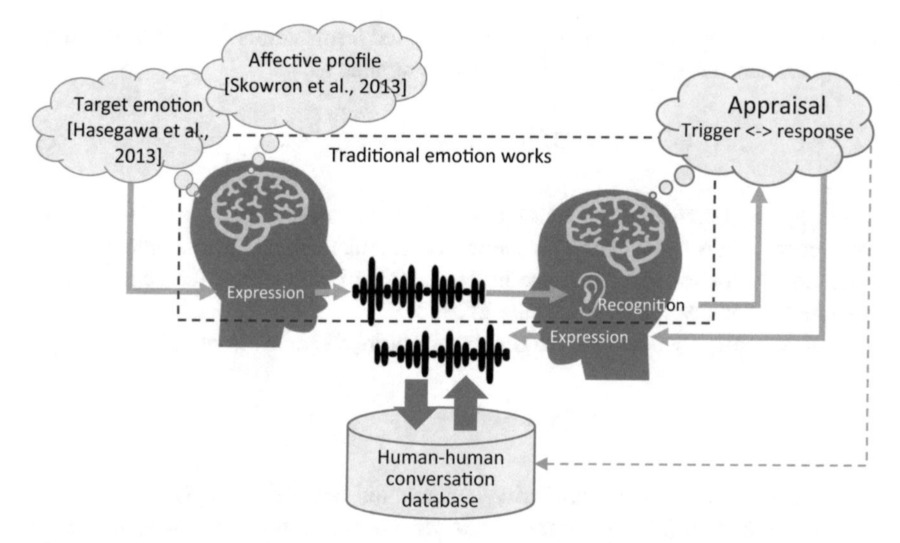

Fig. 1 Overview of proposed approach to elicit a positive emotional change in HCI using examples of appraisal in human dialogue. Relation to existing works is shown

models and the definition of emotion targets by aiming for a general positive affective interaction. The use of data-driven approach rids the need of complex modeling and manual labor. Text-based human subjective evaluation with crowdsourcing shows that in comparison to the traditional response selection method, the proposed one elicits an overall more positive emotional impact, and yields higher coherence as well as emotional connection.

2 Example-Based Dialogue Modeling (EBDM)

EBDM is a data-driven approach of dialogue modeling that uses a semantically indexed corpus of *query-response*[2] pair examples instead of handcrafted rules or probabilistic models [9]. At a given time, the system will return a response of the best example according to a semantic constraint between the query and example queries. This circumvents the challenge of domain identification and switching—a task particularly hard in chat-oriented systems where no specific goal or domain is predefined beforehand. With increasing amount of available conversational data, EBDM offers a straightforward and effective approach for deploying a dialogue system in any domain.

Lasguido et al. have previously examined the utilization of cosine similarity for response retrieval in an example-based dialogue system [8]. In their approach, the

[2]In the context of dialogue system, we will use term *query* to refer to user's input, and *response* to refer to system's output.

similarity is computed between TF-IDF weighted term vectors of the query and the examples. The TF-IDF weight of term t is computed as:

$$\text{TF-IDF}(t, T) = F_{t,T} \log \frac{|T|}{DF_t}, \tag{1}$$

where $F_{t,T}$ is defined as term frequency of term t in a sentence T, and DF_t as total number of sentences that contains the term t, calculated over the example database. Thus, the vector for each sentence in the database is the size of the database term vocabulary, each weighted according to Eq. 1.

Cosine similarity between two sentence vectors S_q and S_e is computed as:

$$\cos_{sim}(S_q, S_e) = \frac{S_q \cdot S_e}{\|S_q\| \, \|S_e\|}. \tag{2}$$

Given a query, this cosine similarity is computed over all example queries in the database and treated as the example pair scores. The response of the example pair with the highest score is then returned to the user as the system's response.

This approach has a number of benefits. First, The TF-IDF weighting allows emphasis of important words. Such quality is desirable in considering emotion in spoken utterances. Second, as this approach does not rely on explicit domain knowledge, it is practically suited for adaptation into an affective dialogue system. Third, the approach is straightforward and highly reproducible. On that account, it serves as the baseline in this study.

3 Emotion Definition

In this work, we define the emotion scope based on the *circumplex model of affect* [11]. Two dimensions of emotion are defined: *valence* and *arousal*. Valence measures the positivity or negativity of emotion; e.g. the feeling of joy is indicated by positive valence while fear is negative. On the other hand, arousal measures the activity of emotion; e.g. depression is low in arousal (passive), while rage is high (active). Figure 2 illustrates the valence-arousal dimension in respect to a number of common emotion terms.

This model describes the perceived form of emotion, and is able to represent both primary and secondary emotion. Furthermore, it is intuitive and easily adaptable and extendable to either discrete or other dimensional emotion definitions. The long established dimension are core to many works in affective computing and potentially provides useful information even at an early stage of research.

Henceforth, based on this scope of emotion, the term *positive emotion* refers to the emotions with positive valence. Respectively, a *positive emotional change* refers to the change of position in the valence-arousal space where the value of valence after the movement is greater than that of before.

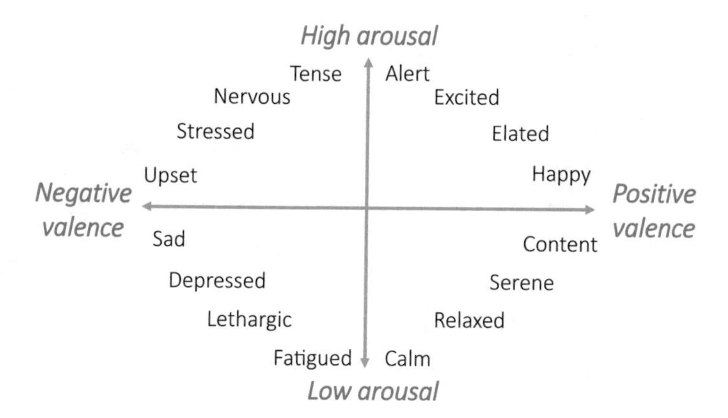

Fig. 2 Emotion dimensions and common terms

4 Proposed Dialogue System

To allow the consideration of the emotional parameters aforementioned, we make use of *tri-turn* units in the selection process in place of the *query-response* pairs in the traditional EBDM approach. A tri-turn consists of three consecutive dialogue turns that are in response to each other, it is previously utilized in collecting *query-response* examples from a text-based conversational data to ensure that an example is dyadic [8]. In this work, we instead exploit the tri-turn format to observe emotional triggers and responses in a conversation.

Within this work, the first, second, and third turns in a tri-turn are referred to as *query*, *response*, and *future*, respectively. The change of emotion observed from *query* to *future* can be regarded as the impact of *response*.

In addition to semantic constraint as described in Sect. 2, we formulate two types of emotional constraints: (1) emotion similarity between the query and the example queries, and (2) expected emotional impact of the candidate responses. Figure 3 illustrates the general idea of the baseline (a) and proposed (b) approaches.

First, we measure *emotion similarity* by computing the Pearson's correlation coefficient of the emotion vector between the query and the example queries, i.e. the first turn of the tri-turns. Correlation r_{qe} between two emotion representation vectors for query q and example e of length n is calculated using Eq. 3,

$$r_{qe} = \frac{\sum_{i=1}^{n}(q_i - \bar{q})(e_i - \bar{e})}{\sqrt{\sum_{i=1}^{n}(q_i - \bar{q})^2}\sqrt{\sum_{i=1}^{n}(e_i - \bar{e})^2}}. \quad (3)$$

This similarity measure utilizes real-time valence-arousal values instead of discrete emotion label. In contrast with discrete label, real-time annotation captures emotion fluctuation within an utterance, represented with the values of valence or arousal with a constant time interval, e.g. a value for every second.

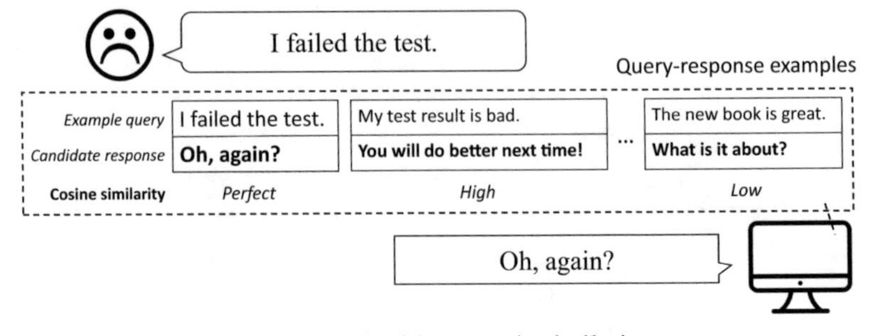

(a) Selection with semantic similarity

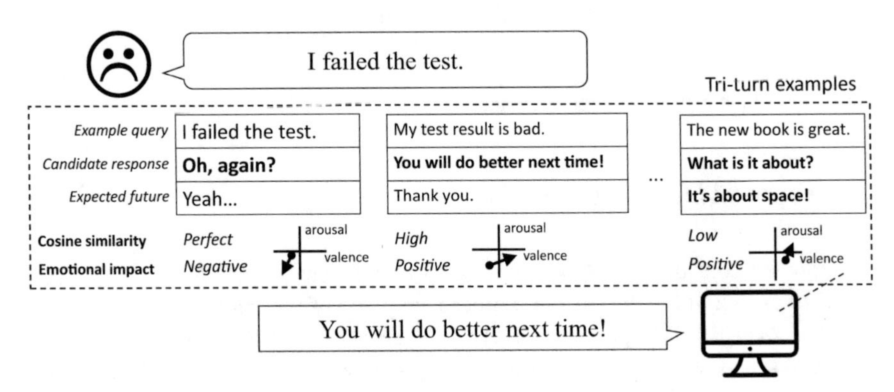

(b) Selection with semantic similarity and emotion parameters

Fig. 3 Response selection in baseline and proposed systems

As the length of emotion vector depends on the duration of the utterance, prior to emotion similarity calculation, sampling is performed to keep the emotion vector in uniform length of n. For shorter utterances with fewer than n values in the emotion vector, we perform sampling with replacement, i.e. a number can be sampled more than once. The sampling preserves distribution of the values in the original emotion vector. We calculate the emotion similarity score separately for valence and arousal, and then take the average as the final score.

Secondly, we measure the *expected emotional impact* of the candidate responses. In a tri-turn, emotional impact of a *response* according to the *query* and *future* is computed using Eq. 4.

$$\text{impact}(response) = \frac{1}{n}\sum_{i=1}^{n} f_i - \frac{1}{n}\sum_{i=1}^{n} q_i, \tag{4}$$

where q and f are the emotion vectors of *query* and *future*. In other words, the actual emotion impact observed in an example is the expected emotional impact during

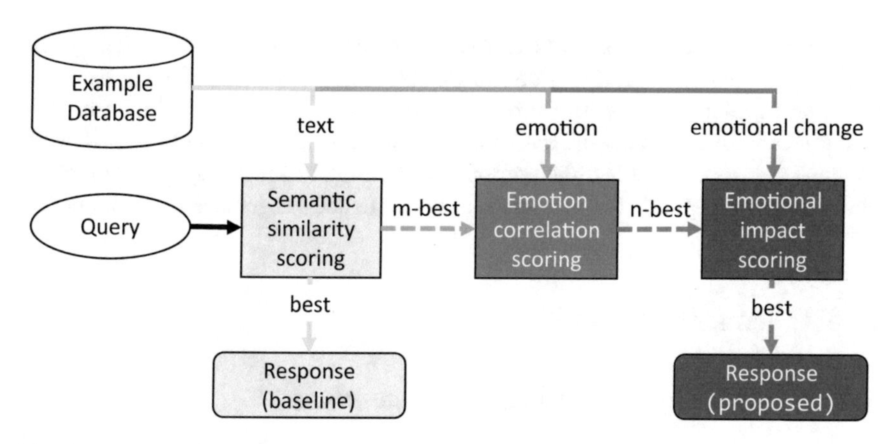

Fig. 4 Steps of response selection

the real interaction. For expected emotional impact, we consider only valence as the final score.

Figure 4 illustrates the steps of response selection of the baseline and proposed systems. We perform the selection in three steps based on the defined constraints. For each step, a new score is calculated and re-ranking is performed only with the new score, i.e. no fusion with the previous score is performed.

The baseline system will output the *response* of the tri-turn example with the highest semantic similarity score (Eq. 2). On the other hand, on the proposed system's response selection, we pass m examples with highest semantic similarity scores to the next step and calculate their emotion similarity scores (Eq. 3). From n examples with highest emotion similarity scores, we output the *response* of the tri-turn example with the most positive expected emotional impact (Eq. 4).

The semantic and emotion similarities are important in ensuring an emotional response that is as close as possible with the example. The two similarity scores are processed incrementally considering the huge difference of their space sizes. The two-dimensional emotion space is much smaller than that of the semantic, making it more likely for emotions to have a high similarity score. Thus, imposing the emotion constraints in the reduced pool of semantically similar examples will help achieve a more relevant result. Furthermore, this reduces the computation time since the number of examples to be scored will be greatly minimized. When working with big example databases, this property is beneficial in giving a timely response.

There are two important points to note regarding the proposed approach. First, the *future* of each tri-turn is not considered as a definite prediction of user response when interacting with the system. Instead, each tri-turn acts as an example of human's appraisal in a conversation with certain semantic and emotional contexts. In real interaction, given similar semantic and emotional contexts with an example tri-turn's *query*, when the system outputs the *response*, the user may experience an emotional change consistent with that of the *future*.

Second, this strategy does not translate to selection of the *response* with the most positive emotion. Instead, it is equivalent to selecting the *response* that has the most potential in eliciting a positive emotional response, regardless of the emotion it actually contains. Even though there is no explicit dialogue strategy to be followed, we expect the data to reflect the appropriate situation to show negative emotion to elicit a positive impact in the user, such as relating to one's anger or showing empathy.

5 Experimental Set up

5.1 Emotionally Colorful Conversational Data

To achieve a more natural conversation, we utilize an emotionally colored corpus of human spoken interaction to build the dialogue system. In this section, we describe in detail the SEMAINE database and highlight the qualities that make it suitable for our study.

The SEMAINE database consists of dialogues between a user and a Sensitive Artificial Listener (SAL) in a Wizard-of-Oz fashion [10]. A SAL is a system capable of holding a multimodal conversation with humans, involving speech, head movements, and facial expressions, topped with emotional coloring [13]. This emotional coloring is adjusted according to each of the SAL characters; cheerful Poppy, angry Spike, sad Obadiah, and sensible Prudence.

The corpus consists of a number of sessions, in which a user is interacting with a wizard SAL character. Each user interacts with all 4 characters, with each interaction typically lasting for 5 min. The topics of conversation are spontaneous, with a limitation that the SAL can not answer any questions.

The emotion occurrences are annotated using the FEELtrace system [3] to allow recording of perceived emotion in real time. As an annotator is watching a target person in a video recording, they would move a cursor along a linear scale on an adjacent window to indicate the perceived emotional aspect (e.g. valence or arousal) of the target. This results in a sequence of real numbers ranging from -1 to 1, called a *trace*, that shows how a certain emotional aspect fall and rise within an interaction. The numbers in a trace are provided with an interval of 0.02 s.

In this study, we consider 66 sessions from the corpus based on transcription and emotion annotation availability; 17 Poppy's sessions, 16 Spike, 17 Obadiah, 16 Prudence. For every dialogue turn, we keep the speaker information, time alignment, transcription, and emotion traces.

5.2 Set up

As will be elaborated in Sect. 6.1, in the SEMAINE Corpus, Poppy and Prudence tend to draw the user into the positive-valence region of emotion as opposed to Spike and Obadiah. This resembles a *positive emotional impact*, where the final emotional state is more positive than the initial. Thus, we exclusively use sessions of Poppy and Prudence to construct the example database.

We partition the recording sessions in the corpus into training and test sets. The training set and test set comprise 29 (15 Poppy, 14 Prudence) and 4 (2 Poppy, 2 Prudence) sessions, respectively. We construct the example database exclusively from the training set, containing 1105 tri-turns.

As described in the emotion definition, in this study, we exclusively observe valence and arousal from the emotion annotation. We average as many annotations as provided in a session to obtain the final emotion label. We sample the emotion trace of every dialogue turn into 100-length vectors to keep the length uniform, as discussed in Sect. 4.

In this phase of the research, we utilize the transcription and emotion annotation provided from the corpus as information of the tri-turns to isolate the errors of automatic speech and emotion recognition. For the n-best filtering, we chose 10 for the semantic similarity constraint and 3 for the emotion considering the size of the corpus.

6 Analysis and Evaluation

6.1 Emotional Impact Analysis

We suspect that the distinct characteristic of each SAL affects the user's emotional state in different ways. To observe the emotional impact of the dialogue turns in the data, we extract tri-turn units from the selected 66 sessions of the corpus. As SEMAINE contains only dyadic interactions, a turn can always be assumed as a response to the previous one.

We investigate whether the characteristics of the SAL affect the tendencies of emotion occurrences in a conversation by analyzing the extracted tri-turns. From all the tri-turns extracted from the subset, we compute the emotional impacts and plot them onto the valence-arousal axes, separated by the SAL to show emotion trends of each one. Figure 5 presents this information.

The figure shows different emotional paths taken during the conversation with distinct trends. In Poppy's and Prudence's sessions, most of the emotional occurrences and transitions happen in positive-valence and positive-arousal region with occasional movement to the negative-valence region. On the other hand, in Spike's sessions, movement to the negative-valence positive-arousal region is significantly more often compared to the others. The same phenomenon occurs with negative-valence negative-arousal region in Obadiah's. This tendency is consistent to the characteristic portrayed by each SAL, as our initial intuition suggests.

6.2 Human Evaluation

We perform an offline subjective evaluation to qualitatively measure perceived differences between the two response selection methods. From the test set, we extract 198 test queries. For each test query, we generate responses using the baseline and proposed systems. Queries with identical responses from the two systems are excluded from the evaluation. We further filter the queries based on utterance length, to give enough context to the evaluators; and emotion labels, to give variance in the evaluation. In the end, 50 queries are selected.

We perform subjective evaluation of the systems with crowdsourcing. The query and responses are presented in form of text. We ask the evaluators to compare the systems' responses in respect to the test queries. For each test query, the responses from the systems are presented with random ordering, and the evaluators are asked three questions, adapted from [14]:

1. Which response is more coherent? Coherence refers to the logical continuity of the dialogue.
2. Which response has more potential in building emotional connection between the speakers? Emotional connection refers to the potential of continued interaction and relationship development.
3. Which response gives a more positive emotional impact? Emotional impact refers to the potential emotional change the response may cause.

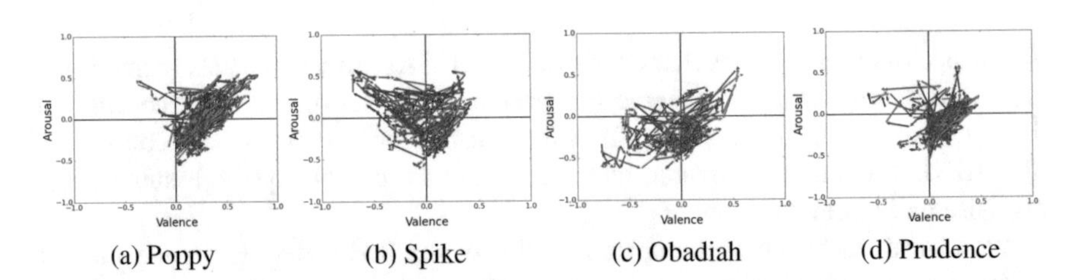

Fig. 5 Emotional changes in SEMAINE sessions separated by SAL character. X- and y-axes show changes in valence and arousal, respectively. Arrows represent emotional impact, with initial emotion as starting point and and final as ending. The direction of the arrows shows the emotional change that occurs. Up and down directions show the increase and decrease of arousal. Right and left directions show the increase and decrease of valence

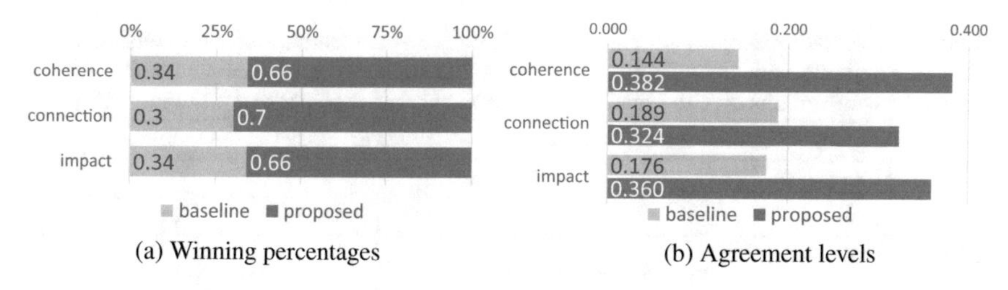

(a) Winning percentages

(b) Agreement levels

Fig. 6 Human evaluation result

50 judgements are collected per query. Each judgment is weighted with the level of trust of the worker.[3] The final judgement of each query for each question is based on the total weight of the overall judgements—the system with the greater weight wins.

Figure 6a presents the winning percentage of each system for each criterion. It is shown that in comparison to the baseline system, the proposed system is perceived as more coherent, having more potential in building emotional connection, and giving a more positive emotional impact.

We investigate this result further by computing the agreement of the final judgement using the Fleiss' Kappa formula. This result is presented in Fig. 6b. We separate the queries based on the winning system and compute the overall agreement of the 50 judgements respectively. It is revealed that the queries where the proposed system wins have far stronger agreement than that where the baseline system wins. Higher agreement level suggests a stronger win, where bigger majority of the evaluators are voting for the winner.

6.3 Discussion

We analyze the consequence of re-ranking and the effect of emotion similarity in the response selection using queries extracted from the test set. Table 1 presents the 10-best semantic similarity ranking, re-ranked and filtered into 3-best emotion similarity ranking, and the candidate response that passed the filtering with the best emotional impact. Note that, as described in Sect. 4: (1) the semantic and emotion similarity scores are computed between the query and example queries (i.e. first turn of the tri-turns), (2) the impact scores are computed from the example queries and example future (i.e. first and third turns of the tri-turns) and (3) the candidate responses are the second turn of the tri-turns. The table shows that the proposed method can select one of the candidate responses that even though is not the best in semantic similarity score, has a higher score in terms of emotion similarity and expected impact.

Furthermore, the proposed selection method is able to generate different responses to identical textual input with different emotional contexts. Table 2 demonstrates this

[3]The level of trust is provided by the crowdsourcing platform we employ in this evaluation. In this evaluation, we employ workers with high-ranking level of trust.

Table 1 Candidate responses re-ranking based on three consecutive selection constraints: (1) semantic similarity with example queries, (2) emotion similarity with example queries, and (3) expected emotional impact of the candidate response. *: baseline response, **: proposed response

Query: Em going to London tomorrow (*valence: 0.39, arousal: −0.11*)

Candidate responses	Ranking steps		
	Semantic	Emotion	Impact
∗And where in Australia?	1		
[laugh]	2		
Organised people need to have holiday	3	1	
It would be very unwise for us to discuss possible external examiners	4		
[laugh]	5		
It's good that sounds eh like a good thing to do, although you wouldn't want to em overspend	6		
That sounds interesting you've quite a lot going on so you need to manage your time	7	2	
Yes	8		
Mhm	9		
**That sounds nice	10	3	1

Table 2 Baseline and proposed responses for identical text with different emotional contexts. The proposed system can adapt to user emotion, while baseline method outputs the same response

Query: Thank you (*valence: 0.13, arousal: −0.18*)	**Query**: Thank you (*valence: 0.43, arousal: 0.05*)
Baseline: Thank you very much that	**Baseline**: Thank you very much that
Proposed: And I hope that everything goes exactly according to plan	**Proposed**: It is always a pleasure talking to you you're just like me

quality. This shows system's ability to adapt to user's emotion in giving a response. These qualities can contribute towards a more pleasant and emotionally positive HCI.

7 Conclusions

We presented a novel attempt in eliciting positive emotional impact in dialogue response selection by utilizing examples of human appraisal in spoken dialogue. We use tri-turn units in place of the traditional query-response pairs to observe emotional triggers and responses in the example database. We augment the response selection criteria to take into account emotion similarity between query and the example query, as well as the expected future impact of the candidate response.

Human subjective evaluation showed that the proposed system can elicit a more positive emotional impact in the user, as well as achieve higher coherence and emotional connection. The data-driven approach we employ in this study is straightforward and could be efficiently replicated and extended. With the increasing access to data and the advancements in emotion recognition, a large unlabeled corpus of conversational data could be used to extensively expand the example database.

In the future, we look forward to try a more sophisticated function for estimating the emotional impact. We hope to test the proposed idea further in a setting closer to real conversation, e.g. by using spontaneous social interactions, considering interaction history, and using real-time emotion recognition. We also hope to apply this idea to a more complex dialogue models, such as Partially Observable Markov Decision Process (POMDP) and to learn an explicit dialogue with machine learning approaches.

Acknowledgements This research and development work was supported by the MIC/SCOPE #152307004.

References

1. Ball G, Breese J (2000) Emotion and personality in a conversational agent. Embodied conversational agents, pp. 189–219
2. Cohen AN, Hammen C, Henry RM, Daley SE (2004) Effects of stress and social support on recurrence in bipolar disorder. Journal of affective disorders 82(1):143–147
3. Cowie R, Douglas-Cowie E, Savvidou S, McMahon E, Sawey M, Schröder M (2000) 'FEELTRACE': an instrument for recording perceived emotion in real time. In: ISCA tutorial and research workshop (ITRW) on speech and emotion
4. Dias J, Mascarenhas S, Paiva A (2014) Fatima modular: towards an agent architecture with a generic appraisal framework. In: Emotion modeling. Springer, pp. 44–56
5. Diener E, Larsen RJ (1993) The experience of emotional well-being
6. Egges A, Kshirsagar S, Magnenat-Thalmann N (2004) Generic personality and emotion simulation for conversational agents. Computer animation and virtual worlds 15(1):1–13
7. Hasegawa T, Kaji N, Yoshinaga N, Toyoda M (2013) Predicting and eliciting addressee's emotion in online dialogue. In: Proceedings of association for computational linguistics, vol 1, pp 964–972
8. Lasguido N, Sakti S, Neubig G, Toda T, Nakamura S (2014) Utilizing human-to-human conversation examples for a multi domain chat-oriented dialog system. Trans Inf Syst 97(6):1497–1505
9. Lee C, Jung S, Kim S, Lee GG (2009) Example-based dialog modeling for practical multi-domain dialog system. Speech Commun 51(5):466–484
10. McKeown G, Valstar M, Cowie R, Pantic M, Schroder M (2012) The SEMAINE database: annotated multimodal records of emotionally colored conversations between a person and a limited agent. Trans Affect Comput 3(1):5–17
11. Russell JA (1980) A circumplex model of affect. J Pers Soc Psychol 39(6):1161
12. Scherer KR, Schorr A, Johnstone T (2001) Appraisal processes in emotion: theory, methods, research. Oxford University Press

13. Schröder M, Bevacqua E, Cowie R, Eyben F, Gunes H, Heylen D, Maat MT, McKeown G, Pammi S, Pantic M et al (2012) Building autonomous sensitive artificial listeners. Trans Affect Comput 3(2):165–183
14. Skowron M, Theunis M, Rank S, Kappas A (2013) Affect and social processes in online communication-experiments with an affective dialog system. Trans Affect Comput 4(3):267–279

Predicting Interaction Quality in Customer Service Dialogs

Svetlana Stoyanchev, Soumi Maiti and Srinivas Bangalore

Abstract In this paper, we apply a dialog evaluation *Interaction Quality (IQ)* framework to human-computer customer service dialogs. IQ framework can be used to predict user satisfaction at an utterance level in a dialog. Such a rating framework is useful for online adaptation of dialog system behavior and increasing user engagement through personalization. We annotated a dataset of 120 human-computer dialogs from two customer service application domains with IQ scores. Our inter-annotator agreement ($\rho = 0.72/0.66$) is similar to the agreement observed on the IQ annotations of publicly available bus information corpus. The IQ prediction performance of an *in-domain* SVM model trained on a small set of call center domain dialogs achieves a correlation of $\rho = 0.53/0.56$ measured against the annotated IQ scores. A *generic* model built exclusively on public LEGO data achieves 94%/65% of the in-domain model's performance. An *adapted* model built by extending a public dataset with a small set of dialogs in a target domain achieves 102%/81% of the in-domain model's performance.

Keywords Human-computer dialog · Interaction quality · User satisfaction

1 Introduction

Automated call/chat centers handle thousands of customer service requests daily and require regular procedures to assess quality of the dialog interaction. While using automation for customer service leads to cost savings, it is important to maintain a high quality of interaction as it affects customers perception of the company and

S. Stoyanchev (✉) · S. Bangalore
Interactions LLC, New Jersey, NJ 07932, USA
e-mail: sstoyanchev@interactions.com; svetlana.stoyanchev@gmail.com

S. Bangalore
e-mail: sbangalore@interactions.com

S. Maiti
CUNY, The Graduate Center, New York, NY, USA
e-mail: smaiti@gradcenter.cuny.edu

M. Eskenazi et al. (eds.), *Advanced Social Interaction with Agents*, Lecture Notes in Electrical Engineering 510, https://doi.org/10.1007/978-3-319-92108-2_16

the brand. Analysts in automated customer support call centers measure objective task success as well as subjective interaction quality. Task success is determined by objective metrics, such as if a customer request was handled appropriately or whether the required information was elicited from the customers. Subjective metrics estimate customers opinion about the quality of the dialog. While an objective task success measure is an important metric for any interface, a subjective dialog evaluation has important long-term implication. A customer's inclination to recommend the system is measured by the Net Promoter Score questionnaires which is widely used by businesses [9].

In this work, we attempt to quantify the subjective user satisfaction of a customer during conversation with an automated spoken customer service dialog system. We apply Interaction Quality framework, first introduced by Schmitt et al. [12]. IQ score, an integer in the range of 1 to 5, is annotated by a labeler on each dialog turn. It has been experimentally shown to correlate with the overall satisfaction of a dialog system user [13]. IQ framework supports prediction of user satisfaction in an ongoing dialog which allows adaptation of dialog behavior to the perceived user satisfaction [16, 17]. In a post-deployment system analysis, IQ prediction on the last turn of a dialog can be used to identify problematic dialogs and infer conditions that lead to decreased user satisfaction [11, 21].

We apply IQ framework in a new domain of customer service dialogs and evaluate generalization of a Support Vector Machine Model trained on the publicly available *LEGO* corpus to the customer service domain. We annotate a dataset of 120 dialogs for two customer service domains (devices and hospitality) using IQ guidelines. Our inter-annotator agreement is similar to the agreement achieved by the annotators of the LEGO corpus, with Weighed Cohen's Kappa $\kappa = 0.54/0.63$ and Spearman's Rank Correlation $\rho = 0.72/0.66$ for each of the domains. We evaluate the automatic IQ prediction using Support Vector Machine Model. The performance of an *in-domain* model trained on a small set of the dialogs in the corresponding call center domain achieves $\rho = 0.53/0.56$. A *generic* model that was built only on public data achieves 94%/65% of the in-domain model's performance. An *adapted* model built by extending a public dataset with a small set of 30 dialogs in a target domain achieves 102%/81% of the in-domain model's performance.

To our knowledge, this is the first application of the IQ framework to a commercially deployed dialog system. Our results indicate that an IQ model trained on a publicly available corpus can be successfully applied and adapted to predict IQ scores in customer service dialogs.

2 Related Work

Roy et al. [10] describe a comprehensive analytics tool for evaluating agent behavior in human call centers. In addition to objective measures of system functions, subjective measures of user satisfaction are also used in evaluation of dialog systems. Subjective measures are evaluated using a questionnaire with a set of subjective

Table 1 Statistics on the LEGO and INTER data sets

Corpus	Num dialogs	Num turns	Avg length	Avg IQ	Avg weighed kappa	Rho
Public *LEGO* dataset						
LEGO1	237	6.3 K	26.9	3.5	0.54	0.72
LEGO2	437	11.1 K	25.4	3.9	0.58	0.72
Proprietary *INTER* dataset						
INTER-D	60	419	6.98	4.3	0.54	0.72
INTER-H	60	393	6.55	4.4	0.62	0.66

questions [4, 5]. Net Promoter Score proposes to simplify the measure to a single question of a hypothetical recommendation [9].

A body of research in dialog focuses on automatic estimation of user perception of the dialog quality using objective dialog measures, such as percentages of `timeout`, `rejection`, `help`, `cancel`, and `barge-in` [1, 20]. Predicting subjective ratings assigned by the actual user requires a set up where users rate the system. However, such an evaluation is not always feasible with real users of commercial systems. Evanini et al. [3] collect labels from external listeners, not the callers themselves, and show high degree of correlation among several human annotators and automatic predictors.

Interaction Quality framework has been validated in user studies showing that IQ scores assigned by expert annotators correlate with user's ratings [12, 13]. Support Vector Machines (SVM) classification of IQ score was shown to be the most effective method for predicting IQ. Sequence methods have also been evaluated but did not outperform SVM [18]. Using a Recurrent Neural Network for prediction of IQ shows promising results [7].

3 Data

In our study we use two data sets: a publicly available *LEGO* dataset and a proprietary *INTER* dataset (see Table 1). *LEGO* dataset consists of the logs and extracted features from the *Let's Go!* bus information dialog system annotated with Interaction Quality [8, 14]. *LEGO1* contains 237 dialogs and is a subset of *LEGO2* which contains 437 dialogs. *INTER* dataset consists of logs from deployed *Interactions LLC* customer support dialog systems implemented with a knowledge-based dialog manager, statistical speech recognition (ASR) and natural language understanding (NLU) components. The Interactions dialog systems use a human-in-the-loop approach: when the NLU's confidence is low, human agent performs NLU by listening to an utterance. The dialog starts with a system's generic '*How may i help you?*' question. Once the reason for the call is established, the system proceeds to collect domain-specific details, including dates, names, or account and phone numbers. Some of the calls are fulfilled within the automatic system while others are forwarded, together with the collected information, to a human agent. In our experiments, we use only the human-computer portion of the customer service dialogs.

From the Interactions data set, we choose two customer support domains: devices (*INTER-D*) and hospitality (*INTER-H*). For each domain, we randomly select 60 dialogs with the lengths ranging between 5 and 10 turns. Two labelers annotated IQ on each dialog turn according to the same guidelines as those used for annotating *LEGO* corpus [14]. Annotations are performed with an in-house implemented iPython notebook interface [6]. We measure agreement using Weighted Cohen's Kappa (κ) and Spearman's Rank Correlation (ρ) [2, 15]. Both of these measures take into account ordinal nature of the scores reducing the discount of disagreements the smaller the difference is between two ratings. To evaluate inter-annotator-agreement, twenty of the *INTER* dialogs in each domain are annotated by both annotators. We observe similar agreement on the *INTER* and *LEGO* datasets with κ between 0.54 and 0.62. and ρ between 0.66 and 0.72.

Cohen's *kappa* is a statistical measure for inter-annotator agreement. It measures agreement between two annotators for categorical items.

$$\kappa = \frac{\rho_o - \rho_e}{1 - \rho_e} \tag{1}$$

where ρ_o is the relative observed agreement between raters and ρ_e is the probability of chance agreement.

Spearman's *rho* is used to measure rank correlation between two variables.

$$\rho = \frac{\sum_i (x_i - \bar{x})(y_i - \bar{y})}{\sqrt{(\sum_i (x_i - \bar{x}))^2 (\sum_i (y_i - \bar{y}))^2}} \tag{2}$$

where x_i and y_i are corresponding ranks and $\bar{x}$, $\bar{y}$ are the mean ranks.

4 Experiments

4.1 Features

The features used in IQ prediction include the features from the automatic speech recognizer (ASR), dialog manager state (DM), user utterance modality, duration, and text/NLU (see Table 2). In our experiments, we use a set of features that may be automatically extracted from system log (AUTO), a subset of the features distributed with the *LEGO* corpus [14]. The AUTO features excluding text, NLU, and dialog

Table 2 Feature set from the *LEGO* corpus. Binary features are marked with?. Generic features also available in the *INTER* corpus are shown in **bold**

Feature set	Features
ASR (utt/total/mean/window)	**ASR success?, ASR failure?, timeout?, reject?, ASR score**, barge-in?
DM (utt/total/mean/window)	**reprompt?, confirm?, acknowledge?**
DM (this utt)	**prompt type (request/ack/confirm), role, loop**, DM-state
Modality	**voice?, dtmf?, unexpected?**
Duration	**words per utt, utt duration, turn number, dialog-duration**
Text/NLU	system-prompt, user-utt, semantic-parse

state are GENERIC and also transferable across domains.[1] We automatically extract the subset of generic features from the *INTER* corpus. We scale all numeric features to have mean 0 and variance 1 using *sklearn.preprocessing.StandardScaler* with default settings.

4.2 Method

In our experiments, we use SVM classification which has been shown to outperform sequence models in the IQ prediction task [19].[2] For the evaluation metrics, we follow [19] and report Unweighed Average Recall (UAR), linearly weighted Cohen's Kappa (w-κ), and Spearman's Rank Correlation ρ.

Unweighted Average Recall is defined as the sum of class wise recalls r_c divided by the number of classes $|C|$.

$$UAR = \frac{1}{|C|}\sum_{c\in C} r_c \tag{3}$$

Recall r_c for each class c is defined as,

$$r_c = \frac{1}{|R_c|}\sum_{i=1}^{R_c} \delta_{h_i r_i} \tag{4}$$

where δ is the Kronecker-delta, h_i and r_i are the ith pair of hypothesis and reference. $|R_c|$ is the total number of ratings per class c.

[1] *barge-in* feature is GENERIC but not recorded in the *INTER* dataset.

[2] We use *linearSVC* from the *sklearn* package with the default parameters.

Table 3 Results of SVM classification on *LEGO* and *INTER* dataset: Unweighed Average Recall (*UAR*), Weighed Cohen's Kappa (w-κ), and Spearman's Rank Correlation (ρ). The last column shows the % of the ρ in the CROSS and the ADAPT conditions in relation to the DOMAIN condition

Features (condition)	Train	Test	UAR	W-κ	ρ	%
Evaluation on *LEGO* data						
AUTO	LEGO1	LEGO1	0.50	0.61	0.78	–
AUTO w/o text	LEGO1	LEGO1	0.49	0.61	0.77	–
GENERIC	LEGO1	LEGO1	0.49	0.61	0.77	–
AUTO	LEGO2	LEGO2	0.47	0.55	0.70	–
AUTO w/o text	LEGO2	LEGO2	0.45	0.52	0.66	–
GENERIC	LEGO2	LEGO2	0.44	0.48	0.61	–
Evaluation on *INTER* data						
GENERIC (DOMAIN)	INTER-D	INTER-D	0.42	0.38	0.53	100
GENERIC (CROSS)	LEGO1	INTER-D	0.42	0.30	0.50	94
GENERIC (ADAPT20)	LEGO1+20%INTER-D	INTER-D	0.38	0.31	0.49	92
GENERIC (ADAPT50)	LEGO1+50%INTER-D	INTER-D	0.36	0.35	0.54	102
GENERIC (DOMAIN)	INTER-H	INTER-H	0.45	0.38	0.57	100
GENERIC (CROSS)	LEGO1	INTER-H	0.40	0.26	0.37	65
GENERIC (ADAPT20)	LEGO1+20%INTER-H	INTER-H	0.38	0.31	0.42	74
GENERIC (ADAPT50)	LEGO1+50%INTER-H	INTER-H	0.41	0.37	0.46	81

For the evaluation on *LEGO* data, we conduct a 10-fold cross validation by splitting the data set into training and test sets on dialog level. For the evaluation on *INTER* data, we consider three types of conditions: CROSS, ADAPT, and DOMAIN. For the CROSS condition, we train the model on *LEGO*1 data and evaluate on all 60 dialogs from the test corpus. For the ADAPT condition, we run a 10-fold validation experiment by selecting part of the *INTER* data for training and interpolating it with *LEGO*1. Each fold is tested on the remaining *INTER* dialogs. For the ADAPT20 condition we use 20% (12) *INTER* dialogs and for the ADAPT50 condition, we use 50% (30) *INTER* dialogs. For the DOMAIN condition, we perform a 10-fold cross-validation on the full 60 dialogs of the *INTER* data.[3]

4.3 Results

Table 3 shows the results of an SVM classifier predicting IQ score on *LEGO* and *INTER* datasets described in Sect. 3. With the AUTO features on *LEGO*1, SVM

[3] We report the results on *INTER* corpus using LEGO1 for training as it achieved higher scores than the models trained on LEGO2.

classifier achieves $UAR = 0.50$, w-$\kappa = 0.61$ and $\rho = 0.78$.[4] Performance of an SVM classifier with the AUTO features on $LEGO2$ is lower than on $LEGO1$: $UAR = 0.47$, w-$\kappa = 0.55$ and $\rho = 0.70$. We observe that removing non-generic features results in a small drop in performance on $LEGO1$ (ρ drops from 0.78 to 0.77). However, the drop is larger on $LEGO2$ when non-generic features are removed (ρ drops from 0.70 to 0.61). $LEGO2$ is a superset of $LEGO1$. Although all $LEGO$ data originates from the same domain of bus information, part of $LEGO2$ was collected at a later time with a potentially different system components affecting homogeneity of features, such as ASR confidence scores or dialog manager's logic. These results are consistent with the previous work that showed that cross-train-testing on $LEGO1$ and ($LEGO2 - LEGO1$) yields a drop in IQ prediction performance in comparison to using training and testing on $LEGO1$.

Next, we evaluate the IQ prediction performance on the customer service dialogs in two domains: devices (INTER-D) and hospitality (INTER-H). In the DOMAIN condition, where the classifier is trained and tested on the data from the same domain, the classification achieves $UAR = 0.42$, w-$\kappa = 0.38$ and $\rho = .053$ on the domain INTER-D and $UAR = 0.45$, w-$\kappa = 0.38$ and $\rho = 0.57$ on the domain H. In the CROSS condition, the classifier achieves $UAR = 0.42$, w-$\kappa = 0.30$ and $\rho = 0.50$ on the domain D and $UAR = 0.40$, w-$\kappa = 0.26$ and $\rho = 0.37$ on the domain INTER-H. The performance for both domains in the CROSS condition is lower than in the DOMAIN condition. Next, we evaluate the ADAPT conditions. Interpolating a model with domain-specific data consistently yields an improvement in w-κ: $ADAPT50 > ADAPT20 > CROSS$ but not in the UAR measure. w-κ and ρ metrics account for the ordinal nature of the IQ class by penalizing less smaller error. Hence the results suggest that with the addition of the in-domain training data, the classification results is *closer* to the human ranking but does not always yield more exact matches of scores.

4.4 Error Analysis

We analyze the errors made by a classifier on the INTER dataset. Figure 1 shows a heat map of true and predicted IQ scores for the CROSS and ADAPT-50 conditions.[5] We observe that the most weight is concentrated on the higher true scores (4, 5) as the dataset is skewed towards the higher scores. In all four heat maps, the weight is concentrated around the diagonal indicating that the prediction error tends to be within a range of $+/-2$ points. However, majority of the weight is not *on* the diagonal reflecting a low UAR which does not take error size into account.

In the CROSS condition the classifier tends to assign lower scores for the turns labeled as 4 and 5, mislabeling them as 3 and 4. This appears to be corrected in the ADAPT-50 condition where the weight shifts closer to the diagonal. For the INTER-H dataset, in the ADAPT50 condition we observe a weight shift towards

[4] This result is a comparable to the result in [19] on $LEGO$ corpus.

[5] The heat map is drawn on a logarithmic scale.

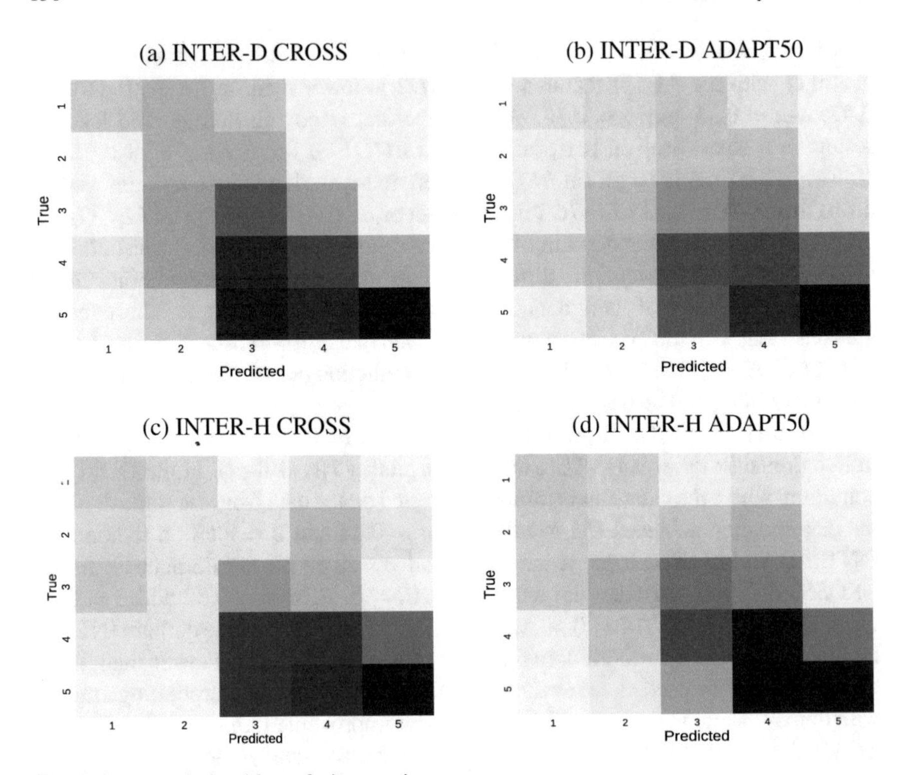

Fig. 1 Error analysis with confusion matrix

the diagonal across all scores. For the INTER-D dataset, however, the lower scores (1, 2) are misclassified more frequently as (3, 4).

We note that the *INTER* dataset is small and highly skewed towards higher scores with the average IQ of 4.2/4.3. Both INTER-D and INTER-H contain very few examples with lower IQ scores. Table 4 shows a confusion matrix with *precision* and *recall* scores for each label for the CROSS condition. More annotated data with lower IQ labels is needed to validate the performance on turns labeled with lower IQ scores.

4.5 Feature Analysis

In this section we explore the relationship between different features and Interaction Quality. We fit a linear regression model using the numeric features of the INTER-D and INTER-H datasets. The weights of the linear model allows us to measure the impact of each feature on IQ prediction. To explore numeric features within INTER dataset, we use all 60 domain specific dialogs to train a linear model for each domain. In Table 5 we list the top 10 numeric features and their weights identified

Table 4 Error analysis confusing matrix for the cross-domain experimental condition CROSS

INTER-D

True(row)/*Pred(col)*	*1*	*2*	*3*	*4*	*5*	**Total**	**Rec**
1	3	5	3	0	0	11	0.27
2	0	5	5	0	0	10	0.5
3	0	7	25	7	0	39	0.64
4	1	7	65	38	6	117	0.32
5	0	3	80	64	95	241	0.39
prec	0.75	0.19	0.14	0.35	0.94	408	0.43

INTER-H

True(row)/*Pred(col)*	*1*	*2*	*3*	*4*	*5*	**Total**	**Rec**
1	3	2	0	1	0	6	0.5
2	2	1	1	4	1	9	0.11
3	3	1	9	5	0	18	0.5
4	1	1	52	56	21	131	0.43
5	0	0	56	69	104	229	0.45
prec	0.33	0.2	0.08	0.41	0.82	387	0.40

by the linear models from each domain. The common top features are marked as **bold**. The feature names with (w) implies the feature was calculated on a window of previous utterances(window size = 3).

We noted that seven features(#reprompt, %ASR success, %ASR Timeout| reject, %ASR reject, Mean ASR score, %reprompt and %unexpected modality) are assigned a top-10 score by a linear model in both datasets. We found #reprompt to be most important feature in INTER-D. The negative sign to weight -5.8245 further denotes that more re-prompt in dialogs results in a lower IQ. %ASR success is the second most important features in INTER-D. This feature is also the most important feature in INTER-H. The feature has positive weight in both dataset indicating that the dialogs with higher %ASR success have higher IQ. We also find the window features are very important for INTER-D. Mean ASR score in last 3 dialogs increase the predicted IQ score. More ASR reject or timeout in previous 3 turns decrease the predicted IQ. For the INTER-H dataset we see that the higher user turn number affects IQ negatively indicating that lower score is more likely to appear later in dialog.

5 Conclusions and Future Work

In this work, we apply *Interaction Quality* dialog evaluation framework to predict user satisfaction in human-computer customer service dialogs. We annotate 120 dialogs from two deployed customer service applications and use them to evaluate IQ prediction performance. The inter-annotator agreement Weighed Cohen's κ of

Table 5 Feature analysis with linear regression

INTER-D		INTER-H	
Weights	Features	Weights	Features
−5.8245	**#reprompt**	+9.6005	**%ASR success**
+5.5439	**%ASR success**	−5.2867	**Mean ASR score**
+2.9553	**%ASR Timeout| reject**	+5.0647	**%ASR Timeout| reject**
+2.9553	**%ASR reject**	+5.0647	**%ASR reject**
+2.1197	Mean ASR score (w)	−4.8677	**#reprompt**
−2.0003	**Mean ASR score**	+2.1534	**%reprompt**
+1.7735	**%reprompt**	−1.6206	user turn number
−1.2341	**% unexpected modality**	−1.5383	unexpected modality(w)
−0.8767	ASR reject (w)	+1.0256	dialog duration
−0.8767	ASR timeout/reject (w)	+0.8635	**% unexpected modality**

0.54/0.62 and Spearman's Rank Correlation ρ of 0.72/0.66 for each of the datasets indicate that IQ model can be applied in customer service domain. The performance of an in-domain model trained only on the call center domain achieves $\rho = 0.53/0.56$. A generic model built only on public data achieves 94%/65% of the in-domain performance. A generic model built by extending a public dataset with a small set of 30 dialogs in a target domain achieves 102%/81% of the in-domain performance. The results of the cross-domain evaluation show that a model built on a publicly available LEGO corpus can be directly applied to customer service dialogs. Further adaptation to the domain yields an improved performance.

In the future work, we will further analyze the features used in predicting IQ and their effect on domain adaptation. We will apply Recurrent Neural Network model to predict *change* in IQ score (UP/SAME/DOWN) using a distributed representation that captures subjectivity of the task and diverging views of the annotators.

References

1. Beringer N, Kartal U, Louka K, Schiel F, Türk U, et al (2002) Promise–a procedure for multimodal interactive system evaluation. In: Multimodal resources and multimodal systems evaluation workshop program, Saturday, June 1, 2002, p 14
2. Cohen J (1968) Weighted kappa: Nominal scale agreement with provision for scaled disagreement or partial credit. Psychological Bulletin 70:213–220
3. Evanini K, Hunter P, Liscombe J, Suendermann D, Dayanidhi K, Pieraccini R (2008) Caller experience: a method for evaluating dialog systems and its automatic prediction. In: Spoken language technology workshop, 2008. SLT 2008. IEEE, pp 129–132
4. Hartikainen M, Salonen EP, Turunen M (2004) Subjective evaluation of spoken dialogue systems using SERVQUAL method. In: INTERSPEECH
5. Hone KS, Graham R (2000) Towards a tool for the subjective assessment of speech system interfaces (SASSI). Nat Lang Eng 6(3–4):287–303

6. Pérez F, Granger BE (2007) IPython: a system for interactive scientific computing. Comput Sci Eng 9(3):21–29
7. Pragst L, Ultes S, Minker W (2017) Recurrent neural network interaction quality estimation. Springer Singapore, Singapore, pp 381–393
8. Raux A, Langner B, Black A, Eskenazi M (2005) Let's Go public! Taking a spoken dialog system to the real world. In: Proceedings of eurospeech
9. Reichheld FF (2004) The one number you need to grow. Harvard business review 81(12):46–54
10. Roy S, Mariappan R, Dandapat S, Srivastava S, Galhotra S, Peddamuthu B (2016) Qart: a system for real-time holistic quality assurance for contact center dialogues. In: Thirtieth AAAI conference on artificial intelligence
11. Schmitt A., Hank C., Liscombe J (2008) Detecting problematic dialogs with automated agents. In: Proceedings of the 4th IEEE tutorial and research workshop on perception and interactive technologies for speech-based systems: perception in multimodal dialogue systems. Springer, Berlin, Heidelberg, pp 72–80
12. Schmitt A, Schatz B, Minker W (2011) Modeling and predicting quality in spoken human-computer interaction. In: Proceedings of the SIGDIAL 2011 conference. Association for Computational Linguistics, pp 173–184
13. Schmitt A, Ultes S (2015) Interaction quality: assessing the quality of ongoing spoken dialog interaction by experts-and how it relates to user satisfaction. Speech Commun 74:12–36
14. Schmitt A, Ultes S, Minker W (2012) A parameterized and annotated spoken dialog corpus of the CMU Let's Go bus information system. In: Proceedings of the eight international conference on language resources and evaluation (LREC'2). European Language Resources Association (ELRA), Istanbul, Turkey
15. Spearman C (1904) The proof and measurement of association between two things. Am J Psychol 15:88–103
16. Suendermann D, Liscombe J, Pieraccini R (2010) Minimally invasive surgery for spoken dialog systems. In: INTERSPEECH, pp 98–101
17. Ultes S, Kraus M, Schmitt A, Minker W (2015) Quality-adaptive spoken dialogue initiative selection and implications on reward modelling. In: Proceedings of the 16th annual meeting of the special interest group on discourse and dialogue. Association for Computational Linguistics, Prague, Czech Republic, pp 374–383
18. Ultes S, Minker W (2014) Interaction quality estimation in spoken dialogue systems using hybrid-HMMs. In: Proceedings of the SIGDIAL 2014 conference, The 15th annual meeting of the special interest group on discourse and dialogue, 18–20 June 2014, Philadelphia, PA, USA, pp 208–217
19. Ultes S, Sánchez MJP, Schmitt A, Minker W (2015) Analysis of an extended interaction quality corpus. In: Natural language dialog systems and intelligent assistants. Springer, pp 41–52
20. Walker M, Kamm C, Litman D (2000) Towards developing general models of usability with paradise. Nat Lang Eng 6(3&4):363–377
21. Walker MA, Langkilde-Geary I, Hastie HW, Wright JH, Gorin A (2002) Automatically training a problematic dialogue predictor for a spoken dialogue system. J Artif Intell Res 16(1):293–319

Direct and Mediated Interaction with a Holocaust Survivor

Bethany Lycan and Ron Artstein

Abstract The *New Dimensions in Testimony* dialogue system was placed in two museums under two distinct conditions: docent-led group interaction, and free interaction with visitors. Analysis of the resulting conversations shows that docent-led interactions have a lower vocabulary and a higher proportion of user utterances that directly relate to the system's subject matter, while free interaction is more personal in nature. Under docent-led interaction the system gives a higher proportion of direct appropriate responses, but overall correct system behavior is about the same in both conditions because the free interaction condition has more instances where the correct system behavior is to avoid a direct response.

1 Introduction

Conversational agents that serve as museum exhibits interact with two populations: museum visitors, and museum docents who may show the system to visitors. Sometimes, docent-led interactions can be useful in initial stages of deployment, before the system is ready for interacting with visitors. For example, the Virtual Museum Guides at the Museum of Science in Boston [2] were initially operated by docents; the system was later extended to enable direct interaction with the public, though even this direct interaction was partly constrained by posting a list of suggested questions, which accounted for 30% of all visitor utterances [3]. But when a system is designed for direct public interaction, several questions arise about the role of museum docents. Is docent-mediated interaction helpful? Does it enhance the visitor experience or detract from it? And how should the needs of museum docents affect the design of the system?

B. Lycan
California State University, Long Beach, CA, USA
e-mail: lycanbg@yahoo.com

R. Artstein (✉)
USC Institute for Creative Technologies, Playa Vista, Los Angeles, CA, USA
e-mail: artstein@ict.usc.edu

M. Eskenazi et al. (eds.), *Advanced Social Interaction with Agents*, Lecture Notes in Electrical Engineering 510, https://doi.org/10.1007/978-3-319-92108-2_17

This paper presents a natural experiment, where the same dialogue system was placed in two museums, under two different conditions—docent-led and open to the public. *New Dimensions in Testimony* [4, 5] is a dialogue system that replicates conversation with Holocaust survivor Pinchas Gutter. Users talk to a persistent representation of Mr. Gutter presented on a large (almost life-size) video screen, and the system selects and plays pre-recorded video clips of the survivor in response to user utterances. The result is much like an ordinary conversation between the user and the survivor. The system was designed from the outset for direct interaction: an extensive testing process resulted in a library of over 1600 video clips, including responses to the most common user questions as well as utterances designed for maintaining coherence and continuity. The system was installed in the Illinois Holocaust Museum and Education Center in Skokie on March 4, 2015, and is still in use as of the time of this writing. Due in part to properties of the physical location and in part to the museum's choice, the exhibit in Illinois is primarily docent-led. Between April 24, 2016 and September 5, 2016, a copy of the system was also installed in the United States Holocaust Memorial Museum in Washington, DC (USHMM). The exhibit at USHMM was built as a booth where individual museum visitors could come up to the system and start a conversation. Comparing the system's operation in the two installations allows us to study both differences in how users interact with the system, and how the system performs in these two distinct settings.

2 Method

The analysis is based on interaction logs from the museums, which contain time-stamped user utterance texts and system response IDs and texts. The user utterance texts are automatic transcriptions by Google Chrome ASR,[1] as logged by the system in real time; previous testing has shown that this ASR has a word error rate of about 5% on this domain [4], so we relied on the ASR output for analysis rather than transcribe the recorded audio. System response IDs identify the video clip used for each response, and the system response texts are the words spoken by Mr. Gutter in the video clip. A sample of the interaction logs from contiguous time periods was selected for quantitative analysis, covering about 2000 user-system interchanges from each museum (Table 1).

While the systems in Illinois and at USHMM are identical, the settings are different. Interaction with *New Dimensions in Testimony* in Illinois is primarily in groups, where communication between visitors and the system is mediated by a museum docent. In a typical interaction the docent will demonstrate a conversation with the survivor, and relay questions from the audience. In contrast, visitors at USHMM talked directly into the microphone connected to the system. Museum docents were available to give background information and offer suggestions in case a visitor needed help, but the docents were specifically instructed to not interfere with the

[1] https://www.google.com/intl/en/chrome/demos/speech.html.

Table 1 Data selected for quantitative analysis

Museum	Dates	User utterances	System responses
Illinois	2015.09.16–2015.11.20	2030	2003
USHMM	2016.05.09–2016.06.08	2025	1995

user's conversation and not make suggestions unless absolutely necessary. In order to encourage natural conversation, users were not provided with any written examples of things they might say to the system. At both museums, the only data recorded were direct audio inputs to the system and the system's action in response; interactions between the docents and the visitors were not recorded. (A separate evaluation observed a sample of interactions and collected user feedback, but these data are not available to us.)

The interaction logs were inspected manually to identify common patterns of interaction. Lexical differences between the user utterances in the two museums were analyzed using the *AntConc* software [1]. User utterances were also annotated by the first author to code consecutive repetitions of (essentially) the same question.

The system's responses were annotated for appropriateness by the first author according to the following scheme. *On-topic responses* are selected by the system when it believes it has found an appropriate response to the user utterance; these were rated on a scale of 1–4, with 1 being irrelevant and 4 being relevant. *Off-topic responses* are utterances used by the system to indicate non-understanding when it is not able to identify a direct response to the user's utterance (for example "please repeat that" or "I don't understand"); these were coded as to whether the decision to use an off-topic was correct (the system does not have a direct response) or an error (the system has a direct response which was not identified).

3 Results

Interactions from USHMM show users relating to Mr. Gutter on a more personal level. Visitors introduce themselves (e.g. *My name is Sheila, I have three granddaughters with me...*), apologize conversationally (*I'm sorry to interrupt you*), and react emotionally to stories told by the survivor (*I'm so sorry to hear that*). In the one instance at USHMM where the survivor asks the visitor why they came to listen to him, the user replies with a long and detailed answer.

The user utterances in Illinois are more tailored to Mr. Gutter's story than those at USHMM. Table 2 shows the most frequent user utterances in the sample from each museum. Many are the same; of those that differ, the questions in Illinois relate more to specific aspects of Mr. Gutter's story, while those asked at USHMM are more interpersonal in nature.

Table 2 Most frequent user utterances (boldfaced utterances appear in only one column)

Illinois		USHMM	
Utterance	N	Utterance	N
Testing	24	Where were you born?	19
Hello	22	How old are you?	17
How old are you?	19	How did you survive?	15
Where were you born?	17	Hello	14
Hello Pincus[a]	16	**Where do you live now?**	14
How are you?	14	Thank you	13
How many languages do you speak?	13	**What happened to your family?**	12
Tell us about your childhood	10	Good morning	10
Can you hear me?	9	**Can you tell us about yourself?**	9
What was life like in the Warsaw Ghetto?	8	How are you today?	8
How did you survive?	8	Do you have any regrets?	8
Do you have any regrets?	8	How are you?	8
Good morning	7	**What's your name?**	8
What was life like before the war?	7	**Hi Pincus**[a]	7
Why didn't the Jews fight back?	7	Hello Pincus[a]	7
How are you today?	6	**Tell me a joke**	6
How did you meet your wife?	6	**What's your favorite color?**	6
Do you have children?	6	**What is your name?**	6
Thank you	6	**Can you tell me a joke?**	6

[a]Mistranscription of Pinchas

A similar observation can be made by looking at the words whose frequency differs the most between the two data sets (Table 3). The top words—*us* in Illinois and *I* at USHMM—reflect the difference between a docent-led group setting and an individual interaction. Other frequent words from Illinois reflect the docents' familiarity with Mr. Gutter's specific story (*Majdanek*, *liberation*, *England*, *Warsaw*), whereas USHMM shows higher relative frequency for *concentration [camp]*, a generic descriptor associated with the Holocaust, as well as words that connect on a personal level (*joke*, *favorite*). Lexical variation is higher at USHMM, with a vocabulary size of 1,386 and density of 10.5 (total tokens divided by vocabulary size), while Illinois has a vocabulary size of 961 and density of 14.2, indicating that docents in Illinois are more likely to stick to familiar topics.

Repetitions of user utterances do not show differences between the two museums. Most repetitions happen after the system gives an inappropriate or off-topic response. In Illinois, instances of user repetition after a seemingly appropriate response give the impression that the docent is trying to elicit a specific utterance they had in mind; however, similar user behavior was observed at USHMM as well.

Table 3 Words which differ the most in frequency between the two corpora; values are the "Keyness" feature from *AntConc* [1]

Illinois							
us	165.9	danik[a]	39.7	war	38.4	liberation	36.7
hear	31.4	what's	30.8	tell	26.9	testing	26.4
didn't	24.9	sing	21.6	England	20.6	why	20.0
life	19.4	about	19.0	after	18.5	Warsaw	17.1
I'm	16.2	rap	16.2	cats	14.7	it's	14.7
USHMM							
I	54.7	it	49.2	concentration	37.8	kosher[b]	27.7
yourself	23.8	don't	23.7	me	23.2	joke	22.2
is	21.2	if	20.1	or	18.1	much	16.9
and	16.8	favorite	16.4	now	16.1	that	15.5
they	13.8	summary	13.7	eat	13.6	experienced	13.2

[a]Mistranscription of *Majdanek*
[b]Mostly from one visitor

Table 4 Appropriateness of responses to user questions

	On-topic responses				Off-topic responses	
	1	2	3	4	OK	Err
Illinois	311	68	65	1346	163	50
USHMM	365	83	99	1169	243	36

The results of the response annotations are shown in Table 4. Illinois has a higher proportion of on-topic responses than USHMM ($\chi^2 = 10$, $df = 1$, $p < 0.005$), which receive overall higher ratings for relevance ($\chi^2 = 24$, $df = 3$, $p < 0.001$). Among the off-topic responses, Illinois has a higher proportion of utterances that should have received a direct response ($\chi^2 = 8.6$, $df = 1$, $p < 0.005$). All of this is expected if the docents in Illinois have a tendency to use familiar utterances—more of these would be recognized, those that are recognized are more likely to lead to an appropriate response, and those that are not recognized are more likely to be misrecognitions by the system rather than questions that cannot be addressed directly. According to these measures, the *New Dimensions in Testimony* system is performing better with the docents, since it yields a higher proportion of appropriate on-topic responses.

However, the difference in performance between the sites is lower if we consider all errors together. Comparing all the appropriate responses (on-topic rated 3–4 and off-topic rated "OK") to the inappropriate ones (on-topic rated 1–2 and off-topic rated "Err"), USHMM data still have a slightly higher proportion of errors (24% compared to 21% at Illinois), but the difference is not highly significant ($\chi^2 = 4.4$, $df = 1$, $p = 0.035$). This is because the higher proportion of off-topic responses at

USHMM represents a correct behavior of the system when faced with user utterances it cannot address directly.

4 Discussion

The main difference between museum visitors and museum docents is familiarity with the dialogue system: docents know the system and have some expectations from it, whereas visitors are likely interacting with the system for the first time. We therefore expect the docents to tailor their utterances to elicit survivor stories they know and like, which is exactly the behavior we observe.

Docents in Illinois are not a passive channel between the visitors and the system, but rather active participants in the dialogue. Do they enhance or detract from the visitor experience? The more interpersonal nature of the USHMM interactions suggests that at least in some respects, a docent-led interaction is inferior, as it interferes with the direct connection between the visitor and the survivor.

Finally, it is interesting to note that overall correct system behavior is about the same in both museums. Shouldn't we expect docent-led interactions, with more limited inputs and better familiarity with the content, to result in better performance? We suspect that the requirements imposed by direct interaction might lead to suboptimal performance in docent-led interaction. In other words: if we had designed the system specifically for use by docents, we probably could have achieved better performance for that population—but worse performance for direct interaction with visitors. A challenge would be to design a dialogue system that could best cater for the needs of both visitors and docents.

Acknowledgements We are grateful to two anonymous reviewers for their comments. The first author was supported by the National Science Foundation under grant 1560426, "REU Site: Research in Interactive Virtual Experiences" (PI: Evan Suma). The second author was supported in part by the U.S. Army; statements and opinions expressed do not necessarily reflect the position or the policy of the United States Government, and no official endorsement should be inferred.

This work is based on materials from *New Dimensions in Testimony*—a collaboration between the USC Institute for Creative Technologies, the USC Shoah Foundation, and Conscience Display, which was made possible by generous donations from private foundations and individuals. We are extremely grateful to The Pears Foundation, Louis F. Smith, and two anonymous donors for their support, and to the Illinois Holocaust Museum and Education Center and the United States Holocaust Memorial Museum for hosting the exhibit. We owe special thanks to Pinchas Gutter for sharing his story, and for his tireless efforts to educate the world about the Holocaust.

References

1. Anthony L (2014) AntConc (version 3.4.3) [computer software]. http://www.laurenceanthony.net/

2. Swartout W, Traum D, Artstein R, Noren D, Debevec P, Bronnenkant K, Williams J, Leuski A, Narayanan S, Piepol D, Lane C, Morie J, Aggarwal P, Liewer M, Chiang JY, Gerten J, Chu S, White K (2010) Ada and Grace: toward realistic and engaging virtual museum guides. In: Allbeck J, Badler N, Bickmore T, Pelachaud C, Safonova A (eds) Intelligent virtual agents: 10th international conference, IVA 2010, Philadelphia, PA, USA, September 20–22, 2010 Proceedings. Lecture notes in artificial intelligence, vol 6356. Springer, Heidelberg, pp 286–300. https://doi.org/10.1007/978-3-642-15892-6_30
3. Traum D, Aggarwal P, Artstein R, Foutz S, Gerten J, Katsamanis A, Leuski A, Noren D, Swartout W (2012) Ada and Grace: direct interaction with museum visitors. In: Nakano Y, Neff M, Paiva A, Walker M (eds) Intelligent virtual agents: 12th international conference, IVA 2012, Santa Cruz, CA, USA, September 12–14, 2012 Proceedings. Lecture notes in artificial intelligence, vol 7502. Springer, Heidelberg, pp 245–251. https://doi.org/10.1007/978-3-642-33197-8_25
4. Traum D, Georgila K, Artstein R, Leuski A (2015a) Evaluating spoken dialogue processing for time-offset interaction. In: Proceedings of the SIGDIAL 2015 conference, Prague, Czech Republic, pp 199–208. http://www.aclweb.org/anthology/W/W15/W15-4629.pdf
5. Traum D, Jones A, Hays K, Maio H, Alexander O, Artstein R, Debevec P, Gainer A, Georgila K, Haase K, Jungblut K, Leuski A, Smith S, Swartout W (2015b) New Dimensions in Testimony: digitally preserving a Holocaust survivor's interactive storytelling. In: Schoenau-Fog H, Bruni LE, Louchart S, Baceviciute S (eds) Interactive storytelling—8th international conference on interactive digital storytelling, ICIDS 2015, Copenhagen, Denmark, November 30–December 4, 2015, Proceedings. Lecture notes in computer science, vol 9445. Springer, Heidelberg, pp 269–281. https://doi.org/10.1007/978-3-319-27036-4_26

Acoustic-Prosodic Entrainment in Multi-party Spoken Dialogues: Does Simple Averaging Extend Existing Pair Measures Properly?

Zahra Rahimi, Diane Litman and Susannah Paletz

Abstract Linguistic entrainment, the tendency of interlocutors to become similar to each other during spoken interaction, is an important characteristic of human speech. Implementing linguistic entrainment in spoken dialogue systems helps to improve the naturalness of the conversation, likability of the agents, and dialogue and task success. The first step toward implementation of such systems is to design proper measures to quantify entrainment. Multi-party entrainment and multi-party spoken dialogue systems have received less attention compared to dyads. In this study, we analyze an existing approach of extending pair measures to team-level entrainment measures, which is based on simple averaging of pairs. We argue that although simple averaging is a good starting point to measure team entrainment, it has several weaknesses in terms of capturing team-specific behaviors specifically related to convergence.

Keywords Entrainment · Multi-party spoken dialogue · Convergence Acoustic-prosodic

1 Introduction

Linguistic entrainment is one of the main characteristics by which conversing humans can improve the naturalness of speech [18]. Linguistic entrainment is the tendency of interlocutors to speak similarly during interactions [3, 21]. Humans entrain to each other in multiple aspects of speech, including acoustic, phonetic, lexical, and syntactic features [13, 19, 20]. Entrainment has been found to be associated with

Z. Rahimi (✉) · D. Litman
University of Pittsburgh, Pittsburgh, PA, USA
e-mail: zar10@pitt.edu

D. Litman
e-mail: dlitman@pitt.edu

S. Paletz
University of Maryland, College Park, MD, USA
e-mail: paletz@umd.edu

M. Eskenazi et al. (eds.), *Advanced Social Interaction with Agents*, Lecture Notes in Electrical Engineering 510, https://doi.org/10.1007/978-3-319-92108-2_18

a variety of conversational qualities and social behaviors, e.g., liking, social attractiveness, positive affect, approval seeking, dialogue success, and task success [2, 10, 20, 22]. Acoustic and lexical entrainment has been implemented in Spoken Dialogue Systems (SDS) in several studies which have shown improvement in rapport, naturalness, and overall performance of the system [12, 15, 16, 18]. Indeed, implementing an entraining SDS is important to improve these systems' performance and quality, measured by user perceptions. All of these SDSes deal with the dyadic interaction of a user and a computer agent, but there are several situations that involve multi-party interaction between an agent and several users or between several agents and users. Foster et al. [6] has studied the interaction of a robot with multiple customers in a dynamic, multi-party social setting. Hiraoka et al. [9] presents an approach for learning dialog policies for multi-party trading dialogs. However, implementation of entrainment in multi-party SDS has not been done yet.

The first step toward the implementation of entrainment in a multi-party SDS, where the agent entrains to the users, is to define proper multi-party entrainment measures. Recently, a few researchers have studied multi-party entrainment in online communities and conversational groups of humans [5, 7, 8, 14]. Regardless of which approach these studies use to measure entrainment, they utilize simple averaging to extend pair-level measures to team-level ones. With a long-term goal of designing more accurate entrainment measures that can demonstrate team-specific characteristics, we perform a qualitative analysis to validate the simple averaging approach. Our hypothesis is that although simple averaging might be a good starting point, it is not capable of capturing several team-specific behaviors. For this purpose, we analyze the behavior of individuals in teams with respect to the team entrainment value. We show that there are at least two team-specific challenges that are not captured well by existing entrainment measures.

Our focus in this paper is on the convergence of acoustic-prosodic features. In the next section we describe the corpus and the convergence measure that we are adopting from prior work. The third section includes the results and discussion of the qualitative within-team analysis with the goal of validating the argument that simple averaging is not a proper approach.

2 Prior Work

In this section, we explain the corpus and the team entrainment measure of convergence that we adapted from prior studies [14].

2.1 The Teams Corpus

The freely available Teams Corpus [14] consists of dialogues of 62 teams of 3–4 participants playing a cooperative board game, Forbidden Island^{TM}. Subjects who were 18 years old or older and native speakers of American English participated in only one session. In each session, participants played the game twice and completed self-reporting surveys about personality characteristics and team cohesion and satisfaction. This game requires cooperation and communication among the players in order to win as a group. The players were video- and audio-taped while playing the game. The corpus consists of 35 three-person and 27 four-person teams. In total, 213 individuals participated in this study, of which 79 are males and 134 are females. In this paper, we only utilized the data from the first game in each session, as it is not necessary to consider cross-game entrainment in order to show evidence for our hypothesis.

2.2 *Multi-party Acoustic Entrainment*

There are two main approaches to quantifying entrainment [11]. The first is a local approach, which focuses on entrainment at a very fine-grained level of adjacent speaking turns [4, 5]. The second is a global approach, which measures entrainment at the conversation level and is the only one considered in this paper. There are several global entrainment measures such as proximity, convergence, and synchrony [13]. Convergence, which is our focus in this paper, measures an increase in similarity of speakers over time. Consistent with the few existing studies measuring multi-party entrainment [5, 7, 14], we use simple averaging of dyad-level measures to build a multi-party measure of entrainment.

To calculate convergence, we choose two disjoint intervals from a conversation and measure the similarity (distance) of speakers on acoustic-prosodic features in each interval. The change in these similarity (distance) values over time indicates the amount of convergence or divergence. The dyad-level difference [13] is the absolute difference between the feature value for a speaker and her partner in each interval. The team difference is the average of these absolute differences for all pairs in the team as defined in Eq. 1 [14]. A significance test on team differences ($TDiff_p$) of the two intervals indicates whether or not significant convergence or divergence has occurred.

$$TDiff_p = \frac{\sum_{\forall i \neq j \in team}(|speaker_i - speaker_j|)}{|team| * (|team| - 1)} \tag{1}$$

Convergence is defined as:

$$Convergence = TDiff_{p,earlierInterval} - TDiff_{p,laterInterval} \tag{2}$$

A positive value is a sign of convergence and a negative value is a sign of divergence. A value of (approximately) zero is a sign of maintenance, indicating that the differences of team members on the specified feature do not change in the two corresponding intervals.

The results in [14] show that, when comparing 9 acoustic-prosodic features over different game 1 intervals (first vs. last 3 min, first vs. last 5 min, first vs. last 7 minutes, and first vs. second half), min pitch and max pitch converge comparing first vs last three minutes of game 1. Shimmer and jitter become more similar (converge) over time on all the examined intervals. The results are validated by constructing artificial versions of the real conversations: for each member of the team, their silence and speech periods within the whole game are randomly permuted. The artificially constructed conversations do not show any sign of convergence on any of the features for any of the examined intervals.

Although the convergence measures were valid, we argue that the measure can be improved by replacing the simple averaging method with a more sophisticated approach. Simple averaging treats all speakers in a team equally and ignores several team-specific characteristics. We argue that not all of the speakers in a team should be treated the same. In the next session, we try to support our argument by a qualitative within-team analysis.

3 Experiments and Discussion

Each game is divided into four equal disjoint intervals to give us better insight into the global behavior of team members, as opposed to selecting two intervals with arbitrary length from the beginning and end of the game. Unlike [14] we do not remove the silences and use the raw audio files, as removing silences may distort the concurrency of our four intervals.

We use Praat [1] to extract the acoustic-prosodic features. Consistent with previous work on dyad entrainment [2, 13, 17], we focus on the features of pitch, intensity, jitter, and shimmer. Pitch describes the frequency, intensity describes the loudness, and jitter and shimmer describe the voice quality by measuring variations of frequency and energy, respectively.

We extract the following 8 acoustic-prosodic features: maximum (max), mean, and standard deviation (SD) of pitch; max, mean, and SD of intensity; local jitter[1]; and local shimmer.[2] The features are extracted from each of the four intervals of each speaker in each team.

First, we perform a significance test to find out which features show significant convergence and on which intervals. The results of the repeated measures of ANOVA with interval as a factor with 4 levels are shown in Table 1. Significant convergence is only observed on shimmer and jitter which were the only features that were found to be significant in the original study, on all intervals examined there which are not the same as here. 'c' and 'd' are indicative of significant convergence on the corresponding intervals. For example, speakers are significantly converging on shimmer from interval 1 to 3.

---

[1]The average absolute difference between the amplitudes of consecutive periods, divided by the average amplitude.

[2]The average absolute difference between consecutive periods, divided by the average amplitude.

Table 1 The results of the repeated measures of ANOVA. * indicates the p-value <0.05. Pairwise comparisons indicate which intervals are significantly different. The direction (convergence or divergence) is represented by *c* and *d* respectively

Features	ANOVA	Pairwise Comparisons					
		1–2	1–3	1–4	2–3	2–4	3–4
Pitch-max							
Pitch-mean	*			d		d	d
Pitch-sd							
Intensity-max							
Intensity-mean							
Intensity-sd							
Shimmer	*		c	c	c	c	
Jitter	*		c	c			

3.1 Is Simple Averaging a Proper Approach?

While previous studies have averaged pairs' entrainment to measure team entrainment, it remains uncertain whether simple averaging is an optimal approach. What are the flaws and weaknesses of this approach and how can we improve them?

We argue that there are some team-specific behaviors that are not properly quantified using the simple averaging method. To demonstrate with real examples from the corpus, we perform a within-team analysis in which we examine the behavior of individuals within each team and the relationship between their behaviors and team convergence.

For this purpose, we draw the plots of raw values of the feature for each team on all 4 intervals. We chose jitter and shimmer as our features of interest since they are the only features that demonstrated significant convergence. We sort all the teams by their convergence values computed by Eq. 2. For example, the convergence of each team from interval 1 to interval 4 is defined as $TDiff_{p,1} - TDiff_{p,4}$.

We examined the plots of all diverging, converging, and maintaining teams. We argue that there are at least two general cases that the simple averaging approach is unable to capture in the teams' behavior. We describe these two cases.

First, how many of the team members are required to converge in order to consider the team to be converging overall? According to the simple averaging method, the answer is that the number of converging or diverging pairs does not matter. As long as the average convergence is higher than the average divergence, we consider the team to be converging. We argue that this answer is not accurate. For example, consider Fig. 1.[3] Each of the plots in this figure shows the values of jitter for each individual in a team over the four intervals. Comparing the first and the last intervals, it appears that Fig. 1b is the most diverging team in the corpus, based on the convergence

[3]We included the gender of speakers in the plots. But, there is no significant effect of gender composition of the teams on convergence value.

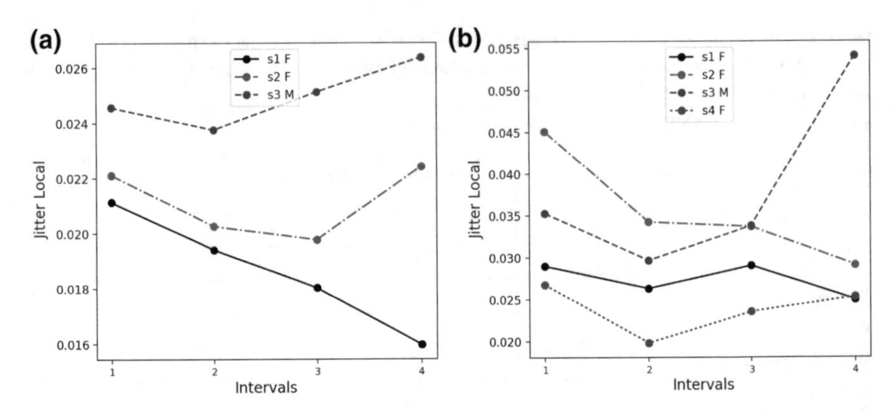

Fig. 1 The plot of jitter of individuals over all four intervals in two diverging teams. Each point is the jitter value calculated for corresponding speaker at corresponding interval using the Praat software. *S* is short for speaker. *M* and *F* are indicative of gender. **a** Convergence from interval 1 to 4 calculated using Eq. 2 is equal to −0.0139. **b** Convergence from interval 1 to 4 calculated using Eq. 2 is equal to−0.0199

measure. But, unlike the team in Fig. 1a, where all the speakers are diverging from each other, speaker 3 is the only participant to diverge from the team, while the rest of the speakers converge. The question is, how much should speaker 3 influence the team convergence in this team?

We argue that the influence of a speaker, such as speaker 3, should lessen if his or her behavior is in the opposite direction of the team behavior. We hypothesize that the solution to this problem is to use a weighted average, where the weights are defined based on the number of speakers that have the same behavior in the team. For example, the weight of a diverging speaker should be the percentage of diverging individuals in the team. This is an ongoing research area. We have seen some improvements using this method, but we have yet to quantify the potential improvement.

Second, does the convergence or divergence of speakers in teams have an absolute meaning, similarly as in pairs? An individual might converge to one teammate while diverging from another one. How do these conflicting behaviors affect the team measure? For example, consider the two teams in Fig. 2. Comparing the first and the last intervals, these two teams have the closest convergence value to zero in our corpus, meaning they are the most maintaining teams. The plot in Fig. 2a is an obvious case of maintenance where none of the team members change their feature values to converge toward or diverge away from the others. But, in Fig. 2b, speaker 1 changes her initial state to converge to speaker 2 while she diverges from speaker 3. We hypothesize that taking into account the self-difference, or how much each speaker's feature value has changed over time, will help to resolve this issue.

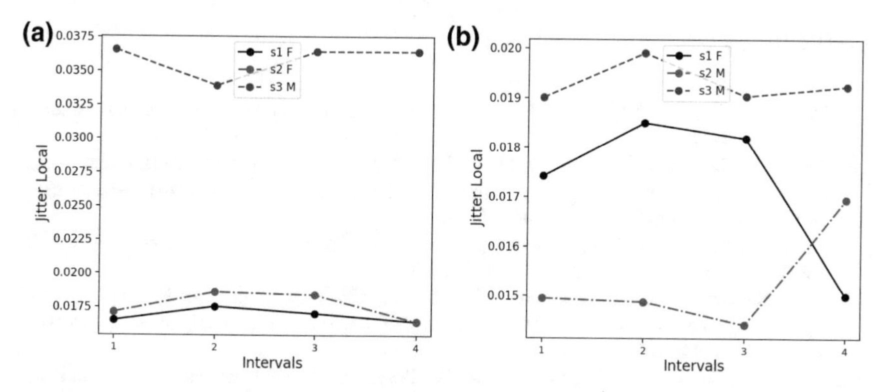

Fig. 2 The plot of jitter of individuals over all four intervals in the two most maintaining teams. Each point is the jitter value calculated for corresponding speaker at corresponding interval using the Praat software. *S* is short for speaker. *M* and *F* are indicative of gender. **a** Convergence from interval 1 to 4 calculated using Eq. 2 is equal to 0.00006. **b** Convergence from interval 1 to 4 calculated using Eq. 2 is equal to −0.00036

4 Conclusion and Future Work

One of the important characteristics of human conversation is linguistic entrainment. Implementing linguistic entrainment in spoken dialogue systems helps to improve the naturalness of the conversation. The first step toward implementation of such systems is to design proper measures to quantify entrainment. Multi-party entrainment and multi-party spoken dialogue systems have gotten less attention compared to dyads. In this study, we analyze the team convergence measure of acoustic-prosodic features in multi-party dialogue. We argue that although existing measures based on simple averaging of pairs are a good starting point to quantify team entrainment, they have several weaknesses in terms of their ability to capture team-specific behaviors. Our study of within-team analyses support our hypothesis. We are currently working on improving the convergence measure by utilizing a weighted average, where the weights are defined based upon the behavior of team members. We will then evaluate the validity of the new measure by comparing its results on real and fake conversations.

Acknowledgements This material is based upon work supported by the National Science Foundation under Grant Nos. 1420784 and 1420377. Any opinions, findings, and conclusions or recommendations expressed in this material are those of the authors and do not necessarily reflect the views of the National Science Foundation.

References

1. Boersma P, Heuven VV (2002) Praat, a system for doing phonetics by computer. Glot Int 5(9/10):341–345
2. Borrie SA, Lubold N, Pon-Barry H (2015) Disordered speech disrupts conversational entrainment: a study of acoustic-prosodic entrainment and communicative success in populations with communication challenges. Front Psychol 6
3. Brennan SE, Clark HH (1996) Conceptual pacts and lexical choice in conversation. J Exp Psychol Learn Memory Cogn 22(6):1482–1493
4. Danescu-Niculescu-Mizil C, Gamon M, Dumais S (2011) Mark my words!: linguistic style accommodation in social media. In: Proceedings of the 20th international conference on world wide web, pp 745–754
5. Danescu-Niculescu-Mizil C, Lee L, Pang B, Kleinberg J (20120 Echoes of power: language effects and power differences in social interaction. In: Proceedings of WWW, pp 699–708
6. Foster ME, Gaschler A, Giuliani M, Isard A, Pateraki M, Petrick RP (2012) Two people walk into a bar: dynamic multi-party social interaction with a robot agent. In: Proceedings of the 14th ACM international conference on multimodal interaction, ICMI'12, pp 3–10
7. Friedberg H, Litman D, Paletz SBF (2012) Lexical entrainment and success in student engineering groups. In: Proceedings fourth IEEE workshop on spoken language technology (SLT). Miami, Florida
8. Gonzales AL, Hancock JT, Pennebaker JW (2009) Language style matching as a predictor of social dynamics in small groups. Commun Res
9. Hiraoka T, Georgila K, Nouri E, Traum D, Nakamura S (2015) Reinforcement learning in multi-party trading dialog. In: 16th annual meeting of the special interest group on discourse and dialogue, p 32
10. Lee CC, Katsamanis A, Black MP, Baucom BR, Georgiou PG, Narayanan S (2011) An analysis of PCA-based vocal entrainment measures in married couples' affective spoken interactions. In: INTERSPEECH, pp 3101–3104
11. Levitan R (2013) Entrainment in spoken dialogue systems: adopting, predicting and influencing user behavior. In: HLT-NAACL, pp 84–90
12. Levitan R, Benuš Š, Gálvez RH, Gravano A, Savoretti F, Trnka M, Weise A, Hirschberg J (2016) Implementing acoustic-prosodic entrainment in a conversational avatar. Interspeech 2016:1166–1170
13. Levitan R, Hirschberg J (2011) Measuring acoustic-prosodic entrainment with respect to multiple levels and dimensions. In: Interspeech
14. Litman D, Paletz S, Rahimi Z, Allegretti S, Rice C (2016)The teams corpus and entrainment in multi-party spoken dialogues. In: 2016 conference on empirical methods in natural language processing (EMNLP). Austin, Texas
15. Lopes J, Eskenazi M, Trancoso I (2013) Automated two-way entrainment to improve spoken dialog system performance. In: ICASSP, pp 8372–8376
16. Lopes J, Eskenazi M, Trancoso I (2015) From rule-based to data-driven lexical entrainment models in spoken dialog systems. Comput Speech Lang 31(1):87–112
17. Lubold N., Pon-Barry H (2014) Acoustic-prosodic entrainment and rapport in collaborative learning dialogues. In: Proceedings of the 2014 ACM workshop on multimodal learning analytics workshop and grand challenge. ACM, pp 5–12
18. Lubold N, Pon-Barry H, Walker E (2015) Naturalness and rapport in a pitch adaptive learning companion. In: 2015 IEEE workshop on automatic speech recognition and understanding (ASRU). IEEE, pp 103–110
19. Mitchell CM, Boyer KE, Lester JC (2012) From strangers to partners: Examining convergence within a longitudinal study of task-oriented dialogue. In: SIGDIAL Conference, pp 94–98
20. Nenkova A, Gravano A, Hirschberg J (2008) High frequency word entrainment in spoken dialogue. In: Proceedings of the 46th annual meeting of the association for computational linguistics on human language technologies: short papers, HLT-Short'08, pp 169–172

21. Porzel R, Scheffler A, Malaka R (2006) How entrainment increases dialogical effectiveness. In: Proceedings of the IUI'06 workshop on effective multimodal dialogue interaction, pp 35–42
22. Reitter D, Moore JD (2007) Predicting success in dialogue. In: Proceedings of the 45th meeting of the association of computational linguistics, pp 808–815

How to Find the Right Voice Commands? Semantic Priming in Task Representations

Patrick Ehrenbrink and Stefan Hillmann

Abstract A common way of interacting with conversational agents is through voice commands. Finding appropriate voice commands can be a challenge for developers and often require empirical methods where participants come up with voice commands for specific tasks. Textual and visual task representations were compared in an online study to assess the influence of semantic priming on spontaneous, user-generated voice commands. Our results indicate that visual representations of tasks lead to a different distribution of used phrases, compared to the textual representation, within the target population. Furthermore, the results suggest that text stimuli can be used to influence user utterances.

Keywords Priming · Task representation · Voice commands

1 Introduction

With the growing importance of mobile devices and the limited usability of input devices that comes with the mobile domain, spoken language becomes more and more important as an input modality for a variety of mobile applications. For the development of a speech-based interface for a given system, it is important to specify an adequate set of voice commands that the system is supposed to understand and act on. Coming up with those commands can be done with a multitude of methods such as interviews, experiments, corpora analysis and so on. A method that is examined more closely in this paper is to run a small exploratory study in which participants solve a task and are asked to formulate proper and intuitive voice commands. Priming can be an intervening factor in this scenario. Here, priming is a psychological effect that functions as a semantic activation. A priming word can activate semantically related words and increases their probability to be used. Furthermore, it facilitates

P. Ehrenbrink (✉) · S. Hillmann
Quality and Usability Lab - Technische Universität Berlin, Berlin, Germany
e-mail: patrick.ehrenbrink@tu-berlin.de

S. Hillmann
e-mail: stefan.hillmann@tu-berlin.de

M. Eskenazi et al. (eds.), *Advanced Social Interaction with Agents*, Lecture Notes in Electrical Engineering 510, https://doi.org/10.1007/978-3-319-92108-2_19

the response to such words by shortening the response time [7]. This means that e.g. if a stimulus like "butter" is presented to a person, then this person will be able to respond faster to "bread" because it is semantically associated with "butter".

Data that was collected in an experiment can be biased if the participants are primed by the written or spoken description of a task. Then, this could lead to less diverse answers and reduce the validity of the results. The effect is especially problematic if we wish to study the frequency distribution of used phrases within the population. Such information can be important for the development of heuristics for automatic speech recognition and natural language understanding, e.g. when resolving ambiguous commands. Especially when interacting with Intelligent Personal Assistants in a spoken dialogue, resolving ambiguity is an important part of achieving a natural conversation. In this paper, we describe an attempt to circumvent the priming of words by using graphical task descriptions instead of textual ones. The two means of task presentation (textual and visual) are compared in the present study in order to determine if they result in a different probability distribution and different absolute quantity of phrases used in the participants' voice commands.

Hypothesis one The textual task description primes the participants so that object names and action words from the task description are more likely to be used in the voice commands that they utter. This will result in a higher proportion of the phrases used in the task description compared to other phrases within the collected user commands. Therefore, we expect that object names and action words from the task description are used more frequently in the responses of the textual condition than in the responses of the visual-only condition.

Hypothesis two When using images instead of text in order to describe tasks, there are no words that could prime the participants. Even though priming is likely to have an effect across different modalities, the cross-modal effect is likely to be smaller compared to the uni-modal effect. With the priming bias for a particular phrase reduced, we expect that this will result in a larger amount of unique phrases compared to the textual condition among the collected user commands.

2 Related Work

There is some previous work that compared priming effects of textual and graphical representations. Bernsen [1] and Dybkjær [4] compared the influence of textual and graphical task descriptions. In one task of their study, the participants had to verbalize a specific time. The specific point in time was indicted either in a textual (e.g. "[…] at 7:20." [4]) or a graphical (picture of a clock) scenario description. They found a priming effect in 74.5% of the answers in the text condition on average. Another study was performed by Möller who used graphical task descriptions in a study about dialogue strategies for spoken dialogue systems [6] for restaurant search. In that study, marked maps were used as task descriptions in order to indicate locations that participants had to search for. Furthermore, shapes of countries were presented to the participant to indicate the nationality of a cuisine (e.g. the shape of Italy was used as an indicator for Italian food). Those graphical task descriptions were used to avoid priming that would possibly have occurred if textual task descriptions had been used. Since no textual task descriptions were used, a comparison between the effects of graphical and textural task descriptions was not performed. Even though those studies present a valid way of representing tasks graphically, the addressed scope in terms of priming is very limited each time. There are not many ways of verbalizing a certain time or indicate a location on a map. Also, the graphics that were used are not (very) ambiguous and leave no room for interpretation. That is, of course, perfectly valid. However, the results are hardly transferable to the task of collecting (a larger number of) suitable utterances that have to be understood by a spoken dialogue system in the addressed domain. From those previous studies, it remains unclear if graphical representations are suitable to gain an overview of intuitive verbalizations of an intended object or action. Such an overview includes the most common names for objects and words for actions that can be expected from the target population together with a quantitative distribution of those verbalizations. The study we present in this paper is aimed to investigate the potential of graphical task descriptions in order to collect information about commonly used phrases for the addressed domain and to avoid priming in user studies with spoken language based systems.

3 Methods and Participants

For this study, we did not use *abstract* graphical representations to describe the task, but photos of actual objects including the surroundings that are to be addressed. Each task is described in two photos in order to indicate the object to be used and the action to be executed. The latter is indicated by a difference between the two images, e.g. a lamp which is off and on. With this approach, we aim at getting a representative overview of verbal commands that our target population would use to trigger an action in a smart home environment. This overview should include names for the objects and descriptions for the actions.

3.1 Study Desgin

An online-study with between-subjects design was performed using the tool Limesurvey 1.90+.[1] The study included two conditions with 13 tasks each. All used images were photos of the devices (i.e. objects) that the participants were supposed to

[1] www.limesurvey.org.

control via a voice command. In the textual condition, that picture was accompanied by a text that stated the task. An example configuration of the stimulus for Task A is shown in Fig. 1. Table 1 shows the descriptions of the 13 tasks that were used in the textual condition and their translation into English. In the visual-only condition, the picture was accompanied by a second picture. That second picture showed the object in a state that was to be induced by a voice command. For example, the first picture showed a lamp that was switched off and the second picture showed a lamp that was switched on. Figure 2 shows the stimulus configuration of Task A in the visual-only condition. Even though other objects such as a table were present in the picture, the object or device in question was always in the center. In the textual condition, the object and action in question were clear to the participant due to the accompanying text. In the visual-only condition, the object and action in question were clear to the participant because both pictures were identical apart from the target object and its functional state. The participants were not really able to control the devices on the pictures. However, they were instructed by a text on top of each questionnaire page

Fig. 1 Screenshot of the stimulus for task A in the text condition

Table 1 English translations of all thirteen German task descriptions

Task	German	English translation
A	Schalten Sie die Tischlampe ein	Switch the table lamp on
B	Schalten Sie die Tischlampe aus	Switch the table lamp off
C	Dimmen Sie die Tischlampe dunkler	Dim the table lamp brighter
D	Dimmen Sie die Tischlampe heller	Dim the table lamp darker
E	Schalten Sie den Ventilator ein	Switch the fan on
F	Schalten Sie den Ventilator aus	Switch the fan off
G	Öffnen Sie das Fenster	Open the window
H	Schließen Sie das Fenster	Close the window
I	Schließen Sie das Rollo	Close the roller blind
J	Öffnen Sie das Rollo	Open the roller blind
K	Machen Sie das Rollo halb zu	Close the roller blind to the half
L	Machen Sie das Rollo halb auf	Open the roller blind to the half
M	Fahren Sie das Rollo ganz herunter	Completely close the roller blind

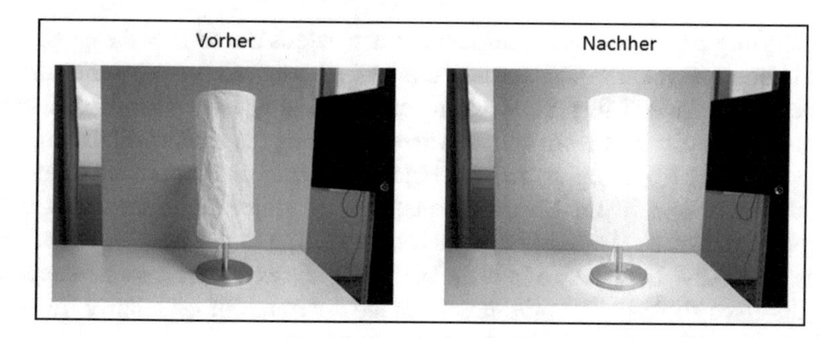

Fig. 2 Screenshot of the stimulus of task A in the visual only condition. Text and images were shown to the participant simultaneously. Translations: *Vorher*: before, *Nachher*: after

that they should phrase a verbal command for a spoken dialogue system that would fulfill the task. Furthermore, they were asked to write the phrased command into a text box at the bottom of the page, as the spoken utterance could not be recorded with the used questionnaire system.

3.2 Participants

In total, 178 persons took part in the survey. Their average age was 31.48 years and the gender was not taken into account. As a compensation for their participation, three vouchers with a value of 40, 20 and 10 Euro were drawn between all participants. Due to technical limitations of the survey tool, participants could not be assigned randomly to a condition. For that reason, they were assigned to the condition according to their respective age.

The participant's age in years was requested by the questionnaire system at the beginning of the trial. If the age was an even number, the participant was assigned to the visual-only condition. In contrast, if the age was an odd number, the textual condition was assigned. This procedure resulted in a total number of 92 participants for the visual-only condition and 86 participants for the textual condition.

3.3 Data Processing

All responses were examined by three different persons (all were members of our research group). They assessed if the requests were appropriate answers to the stated task. All responses that indicated non-compliance or a misunderstanding of the task were excluded from further analysis. Afterwards, the remaining 166 responses (out of 178) were normalized and processed further. Normalization included the removal of all punctuation marks and typing errors as well as the transformation of uppercase characters into lower case. In the subsequent step, words that refer to the name of the object in question (nouns) and words that refer to the actions (verbs) were extracted from each command. This pre-processing ensured that typos or grammatical errors did not result in different word counts when computing the amount of priming.

The amount of priming p_o for the object in a task of the text condition was calculated as shown in Eq. 1. o is the name of the object that was stated in the description of respective task in the textual condition, e.g. "table lamp". Furthermore, $f(o)$ is the relative frequency of o in the set of all phrases that were used to refer to o in the task (see Table 4). The priming of o is the difference of the relative frequency of o in the textual (t) and the visual only (v) condition.

$$p_o = f(o_t) - f(o_v) \tag{1}$$

Equation 2 shows the computation of the amount of priming for the action (a) to be used in a certain task, e.g. "switch on". The annotation is analogue to Eq. 1.

$$p_a = f(a_t) - f(a_v) \tag{2}$$

4 Results

Independent sample t-tests were performed on the frequencies of unique phrases for objects and actions in both conditions with IBM SPSS Statistics 22. The frequencies are provided in Table 2.

Table 2 Frequency of unique commands (i.e. unique utterances) as well as unique phrases (i.e. names) for the object and the action in each task

Task	Commands		Object names		Actions	
	Textual	Visual	Textual	Visual	Textual	Visual
A	26	19	5	5	10	11
B	20	16	5	6	8	7
C	40	35	8	6	23	24
D	37	39	9	6	20	29
E	25	24	7	7	9	10
F	17	24	6	9	7	10
G	17	15	2	4	8	6
H	16	15	1	4	12	7
I	24	39	5	7	13	23
J	22	41	4	8	10	17
K	40	63	3	10	30	41
L	37	57	4	8	32	36
M	29	44	3	8	22	26

In Hypothesis one it was stated that we expect that object names and action phrases that are used in the textual descriptions are more likely to be used in the responses from the textual condition, compared to the responses of the visual-only condition. Results show that the proportion of phrases, used by the participants, that appeared in the textual task description was significantly higher in the textual condition than in the visual-only condition ($\alpha = 0.05$ and $p < 0.001$). Also, the proportion of uttered action phrases from the textual task description was significantly larger in the textual condition than in the visual-only condition ($\alpha = 0.05$ and $p < 0.001$). Table 4 provides the proportions of names and actions from the textual task descriptions as they appeared in each task and condition.

For the comparison of the responses in the two conditions, independent-samples t-tests were performed. The average number of unique object names that were retrieved in each task in the visual-only condition (6.77) was significantly ($\alpha = 0.05$ and $p = 0.024$) larger than the average number of names that were retrieved in each task of the textual condition (4.77). Furthermore, the average number of actions that were retrieved in each task of the visual-only condition (19) was significantly larger ($\alpha = 0.05$ and $p = 0.03$) than the average number of actions that were retrieved in

Table 3 Average numbers of different names and actions for both conditions

Condition	Names	Actions
Textual	4.77	15.69
Visual	6.77	19.00

Table 4 Relative frequency (f) of use of phrases from the textual task description, addressing the object (o) and action (a). Data are shown for the textual (t) and visual only (v) condition. Priming is the difference between the usage in the textual and visual only condition (see Eqs. 1 and 2)

Task	Textual		Visual		Priming	
	$f(o_t)$	$f(a_t)$	$f(o_v)$	$f(a_v)$	p_o	p_a
A	0.202	0.179	0.026	0.052	0.176	0.127
B	0.202	0.905	0.025	0.886	0.177	0.019
C	0.190	0.667	0.026	0.641	0.165	0.026
D	0.179	0.690	0.026	0.487	0.153	0.203
E	0.607	0.226	0.603	0.064	0.005	0.162
F	0.614	0.928	0.622	0.817	−0.007	0.111
G	0.843	0.265	0.789	0.184	0.054	0.081
H	0.843	0.241	0.789	0.263	0.054	−0.022
I	0.821	0.238	0.342	0.132	0.479	0.107
J	0.833	0.238	0.368	0.158	0.465	0.080
K	0.843	0.373	0.297	0.243	0.546	0.130
L	0.845	0.429	0.320	0.293	0.525	0.135
M	0.857	0.369	0.324	0.378	0.533	−0.009

the textual condition (15.69). The average number of object names and actions for each task in both conditions can be seen in Table 2. Table 3 shows the named average frequencies of unique object names and actions.

The amount of priming was calculated according to Eqs. 1 and 2. Priming increased the probability of a name from the task description to be used in a verbal command by 0.25. For the words that describe actions, the average probability increased by 0.08. Besides the amount of priming per task, Table 4 also shows the relative frequencies which were used for the computation of p_o and p_a.

5 Discussion

Hypothesis one can be confirmed. The results show that actions and names for objects that were used in the textual task description appeared significantly more frequently in the responses of the textual condition, compared to the responses of the visual condition. This result can be explained with a priming effect: The task description influenced the likelihood that those words were used by the participants. It can, therefore, be concluded that a textual task description is not optimal for accessing the proportional distribution of individual phrases in a given population and that visual task descriptions should be used if feasible.

Hypothesis two can be confirmed, as well. Our results show that visual task descriptions resulted in the appearance of significantly more names for the objects and actions. It can be concluded, that visual task descriptions are more suitable to collect a wide range of possible utterances than textual descriptions.

From the data, it is evident that the amount of priming is relatively low for action words. One possible explanation for this is the fact that the variety of action words in the tasks that were used is larger compared to the variety of names for the objects. However, this is probably not a sufficient explanation. Additionally, a ceiling effect might have occurred. The large quantity of different action words is mostly the result of expressions that appeared only one or two times. For example, in task B, the majority of participants of both conditions used the word "aus" (engl. "off") (textual: 85%, visual: 84%). Whereas other words such as "ausmachen" (engl. "turn off") or "ausschalten" (engl. "switch off") appeared far less often. The word "aus" was also used in the textual task descriptions. The high usage rate of the word indicates that this is by far the most intuitive action word to be used in that particular situation. Another possible explanation is that the word "aus" can simply be spoken relatively fast and might simply be the most efficient way to perform the task of switching something off.

Priming can appear across different modalities [8]. This means that also a visual or auditory stimulus is able to induce verbal priming. The task presentation probably primed the participant in both conditions. Therefore, the priming effect caused by the text was not this powerful here, but we do not consider this as a problem for the validity of the presented study. If it actually has any effect, it would lessen the observed differences between the conditions. Words that are associated with an object are also primed by its visual appearance, so they are primed by the pictures in this study.

It should be kept in mind that priming from cues, regardless of their modality is what designers can benefit from. Priming can help the users to select the appropriate words for their commands. However, if one wants to examine the distribution of different words in the user population, priming becomes a problem. The words that are used in the task description can then bias the results. For instance, this is an issue, if the intuitive use of a speech-based human-machine interface is to be evaluated.

Besides that obvious impact of priming in the tasks that were used for this study, priming is worth to be considered in a variety of other contexts. One of which is persuasive technology. Since semantic priming represents a cognitive bias towards the primed word or object, this effect can be used to influence people's behavior to a small extent. An advantage is that the priming effect takes place rather subliminally [5] and might, therefore, be an adequate method to avoid negative side effects, such as Psychological Reactance [2, 3], which would otherwise cause the user to be less open to persuasive attempts or even counteract those.

6 Conclusion

Results of a survey that uses visual task representation provide a more realistic overview of the variety and quantity of preferred commands among the target population. This means that obtaining verbal commands by using graphical task descriptions also results in more valid data. Priming effects should be considered when collecting supposedly intuitive commands by running an exploratory study such as the one described in this paper. Apart from being a factor in collecting user utterances, the effect of textual priming could be used to hint users towards correct commands.

Acknowledgements This work was supported by the Bundesministerium für Energie und Wirtschaft (Germany) under grant no. 01MG13001G, Universal Home Control Interfaces @Connected Usability (UHCI).

References

1. Bernsen NO, Dybkjaer H, Dybkjaer L (1998) Designing interactive speech systems. From first ideas to user testing. Springer, London
2. Brehm JW (1966) A theory of psychological reactance. Academic Press, New York
3. Dillard JP, Shen L (2005) On the nature of reactance and its role in persuasive health communication. Commun Monogr 72(2):144–168. https://doi.org/10.1080/03637750500111815
4. Dybkjaer L., Bernsen NO, Dybkjaer H (1994) Scenario design for spoken language dialogue systems development. In: Proceeding of ESCA workshop on spoken dialogue systems, pp 93–96

5. Karremans JC, Stroebe W, Claus J (2006) Beyond vicary's fantasies: the impact of subliminal priming and brand choice. J Exp Soc Psychol 42(6):792–798
6. Möller S (2005) Quality of telephone-based spoken dialogue systems. Kluwer Academic Publishers, Boston
7. Städtler T (2003) Lexikon der Psychologie. Kroner, Stuttgart
8. Swinney DA, Onifer W, Prather P, Hirshkowitz M (1979) Semantic facilitation across sensory modalities in the processing of individual words and sentences. Mem Cogn 7(3):159–165

Exploring the Applicability of Elaborateness and Indirectness in Dialogue Management

Louisa Pragst, Wolfgang Minker and Stefan Ultes

Abstract In this paper, we investigate the applicability of soft changes to system behaviour, namely changing the amount of elaborateness and indirectness displayed. To this end, we examine the impact of elaborateness and indirectness on the perception of human-computer communication in a user study. Here, we show that elaborateness and indirectness influence the user's impression of a dialogue and discuss the implications of our results for adaptive dialogue management. We conclude that elaborateness and indirectness offer valuable possibilities for adaptation and should be incorporated in adaptive dialogue management.

Keywords Communication style · Dialogue management · User study

1 Introduction

Spoken dialogue systems are employed under a number of varying conditions (e.g. the amount of ambient noise or different user states). The user experience when interacting with such a system may be improved if those conditions are taken into account, e.g. by the Dialogue Manager (DM).

In a dialogue system, the DM is the component responsible for keeping the dialogue state updated and selecting the system's next action. There are several DM architectures that enable adaptivity (e.g. [3, 5, 9, 10]), often employing hard adaptation approaches. Such approaches are characterised by utilising dedicated system actions to deal with certain conditions, such as reacting to a angry user by asking what is wrong. Gnjatović et al. [5], for example, present a DM that adjusts the kind

L. Pragst (✉) · W. Minker
Ulm University, Ulm, Germany
e-mail: louisa.pragst@uni-ulm.de

W. Minker
e-mail: wolfgang.minker@uni-ulm.de

S. Ultes
Cambridge University, Cambridge, UK
e-mail: su259@cam.ac.uk

M. Eskenazi et al. (eds.), *Advanced Social Interaction with Agents*, Lecture Notes in Electrical Engineering 510, https://doi.org/10.1007/978-3-319-92108-2_20

of support given to the user to their emotional state, and an in-car dialogue system implemented by Kousidis et al. [7] adapts to situations that require the full attention of the driver by pausing the conversation entirely. Saerbeck et al. [8] vary the social supportiveness of a robotic tutor by using motivational sentences such as 'It was not easy for me either'. Similarly, user studies on adaptive dialogue management often focus on hard adaptations: Jaksic et al. [6] employ fixed phrases as response to the user's emotion, effectively leaving two ways to react: apologising and acknowledging positive emotion. Bertrand et al. [2] offer four kinds of emotional feedback: thankfulness, praise, calm and motivate.

In contrast, soft adaptation is only rarely considered. It is characterised by keeping the propositional content of a system action and changing only the way it is presented, e.g. by phrasing statements more politely if the user is angry. Employing soft changes can offer benefits: it is less obvious to the user than a hard adaptation as it is not directly addressing the cause of the adaptation. Asking the user what is wrong will notify the user of the fact that the dialogue system is reacting to their mood. Continuing with the intended statement, just more politely, can appear a seamless continuation of the dialogue. Additionally, hard adaptations are more likely to disrupt the conversation flow by bringing up a new topic (in this case the emotional state of the user). Finally, hard adaptations can become repetitive if employed too often. Asking the user what is wrong is a viable behaviour once anger is detected, but if the user stays angry it is not advisable to repeat this action within the next few exchanges. In contrast, a soft adaptation may be employed over the course of several exchanges as it can be applied to different statements. Being more polite does not become repetitive if the propositional content of the system actions changes. André et al. [1] have proposed the adjustment of the politeness level for adaptation. While the robotic tutor of Saerbeck et al. [8] uses hard adaptation in the form of fixed motivational sentences, it also employs soft adaptation, e.g. on the wording level by using either 'you' or 'we' to reflect the level of closeness between tutor and student. In this paper, we investigate the applicability of two further means of soft adaptation: the level of elaborateness and indirectness (EI).

Elaborateness refers to the amount of additional information provided to the user. A high level of elaborateness could for example result in giving a weather forecast for the next few days when the user asks for the current weather, while with a low elaborateness only the requested information would be given.

The level of indirectness describes how concretely information is addressed by a speaker. For instance, a direct response to a user request about the current weather would be an accurate description of the weather, e.g., 'It is raining'. An indirect response would be the advice to take an umbrella. In the latter case, the weather is not mentioned directly but can be inferred from the given information.

In the following, we show that EI should be incorporated in adaptive dialogue management as they offer valuable possibilities for adaptation. To this end, we present a user study that answers the following questions: does the perception of a dialogue change with different EI-levels and is the effect of the EI-level dependent on the current situation of the user.

(a) The virtual avatar [4] that was used in our user study.

(b) Louisa depicted as involved in the conversation.

(c) Louisa depicted as distracted from the conversation.

Fig. 1 Screenshots of the videos used in the user studies

The remainder of the paper is structured as follows: Sect. 2 describes the setup and results of our user study. The implications for adaptive dialogue management are discussed in Sect. 3. Finally, we draw our conclusion in Sect. 4.

2 User Study

The goal of our study is to assess the effect of EI on the perception of a dialogue under different conditions. To this end, we utilise two semantically similar dialogues that differ with regard to the level of EI used by the dialogue system. If the assessment of the two dialogues differs, this supports the assumption that EI can be used in adaptive dialogue management.

2.1 Study Design

Our study aims to answer two research questions: does the assessment of a dialogue change with different EI-levels and is the effect of the EI-level dependent on the current situation of the user. This results in two independent variables: the *EI-level* of the dialogue, instantiated with the two levels *low* and *high*, and the *situation* of the user, with the levels *involved* and *distracted*. Differences between the EI-levels signify that participants are receptive to the changes resulting from changing the level of EI and have a preference. If that is the case, EI should be considered when designing the system behaviour with this preference in mind. Furthermore, if the assessment of EI-levels changes in different situations, there is no fixed preference for one specific EI-level and, therefore, EI can be used for adaptation.

Four conditions result from the independent variables: high EI/involved user, low EI/involved user, high EI/distracted user and low EI/distracted user. For each condition, a video showing a dialogue between a human and a virtual agent was recorded.[1,2,3,4] In the videos, a caregiver, called Louisa, interacts with the dialogue system by a virtual avatar called Christian. She suspects that her patient Mr. Smith does not drink enough and expects help from the dialogue system. Screenshots of Christian and Louisa as depicted in the videos can be found in Fig. 1. While our videos presented Christian as a real dialogue system, his behaviour followed scripts that were handcrafted for this study. No actual dialogue system was used in the recordings of the videos. The dialogues in the videos differ slightly between the two EI-levels: Christian uses either a high or a low level of EI. A transcription of the dialogues is provided in the Appendix. Additionally, Louisa is depicted as either involved or distracted in the videos, depending on the situation.

Participants of our study watched two videos, one for each level of EI. The order in which the videos were presented to the participants was randomized. Each participant was assigned to either the *involved* or the *distracted* situation by chance.

We provided our participants with a questionnaire containing ten questions that can be rated using a five-point scale:

Q01: Is Christian helpful?
Q02: Is Louisa emotionally involved in the dialogue?
Q03: Does Christian plan his answers?
Q04: How responsive is Christian to Louisa's contributions?
Q05: Are Christian's answers spontaneous?
Q06: Is Christian emotionally involved in the dialogue?
Q07: How natural is the course of dialogue?
Q08: How much would you like to participate in such a dialogue?
Q09: Which dialogue is more natural?
Q10: In which conversation would you rather participate?

Q01–Q08 were asked for each the *high EI-level* and the *low EI-level* video. Those questions can be rated on a scale from 1–*not at all* to 5–*very much*. Q09 and Q10 compare the videos directly with each other and their scale is labelled from 1–*high EI-level* video to 5–*low EI-level* video.

2.2 Results

Our results show that both of our research question can be answered in the affirmative. Multiple significant differences between the two EI-levels can be found, and the assessment of EI-levels changes depending on the situation. Also, there are significant interaction effects between EI-level and situation. A graphical representation

[1]https://youtu.be/NRDIJ7omlEI.
[2]https://youtu.be/VqXgxFbh5-w.
[3]https://youtu.be/3-ANJmmeZY.
[4]https://youtu.be/bAXbt65vxjw.

of the results can be found in Fig. 2. In the following we present a more detailed description of the results.

Overall, 270 participants took part in our study. 154 participants filled in the questionnaire for the *involved* user and 116 participants reported on the dialogue with the *distracted* user.

We test our hypotheses using T-test and ANOVA. Normal distribution of the data is assumed under the central limit theorem and homogeneity of variances is confirmed by Levene's test.

Slightly more women than men participated in the *involved* condition of the study (61 men, 87 women), and almost twice as many men as women in the *distracted* condition (74 men, 40 women). No significant effect of gender on the results of the study could be found. Most participants were between 20 and 29 years old (*involved*: 84.4%, *distracted*: 72.4%) and had only a limited amount of experience regarding dialogue systems (One time use or less: *involved*: 80.4%, *distracted*: 60.4%).

In all conditions, Christian is perceived as helpful. Changing the EI level does not negatively affect the ability of a dialogue system to help the user with their requests. All ratings for Q01 are significantly higher than 3 ($p < 0.001$).

Furthermore the actor portraying Louisa was perceived as involved in the *involved* condition, achieving a rating significantly higher than 3 for Q02 ($p < 0.001$). In the *distracted* condition, Louisa did not succeed to appear sufficiently distracted, still receiving a rating significantly higher than 3 ($p < 0.001$). However, she at least appears less involved than in the *involved* condition, scoring a significantly lower rating ($p < 0.001$).

When Louisa is depicted as *involved*, participants rate Christian as more spontaneous, responsive, and emotionally involved if he uses a *high EI-level* and more planning ahead if he uses a *low EI-level*. Furthermore, the dialogue is perceived as more natural if a *high EI-level* is used, and participants would rather partake in such a dialogue. Significant differences can be detected for Q03–Q10 ($p < 0.001$ in all cases). Those findings support the hypothesis that the EI-level has a strong impact on the user's perception of the dialogue partner as well as the overall dialogue. Furthermore, this perception changes as Louisa is perceived as less involved in the *distracted* condition. The EI-level no longer influences how responsive, spontaneous and emotionally involved Christian appears and how much participants would like to participate in the dialogue. Q04, Q05, Q06 and Q08 no longer show significant differences between the EI-level in the *distracted* condition of the study, and for Q04 and Q05, a significant interaction effect ($p < 0.001$) can be found between EI-level and situation. This supports our second hypothesis that the effect of EI on the assessment of a dialogue changes in relation to the user's situation.

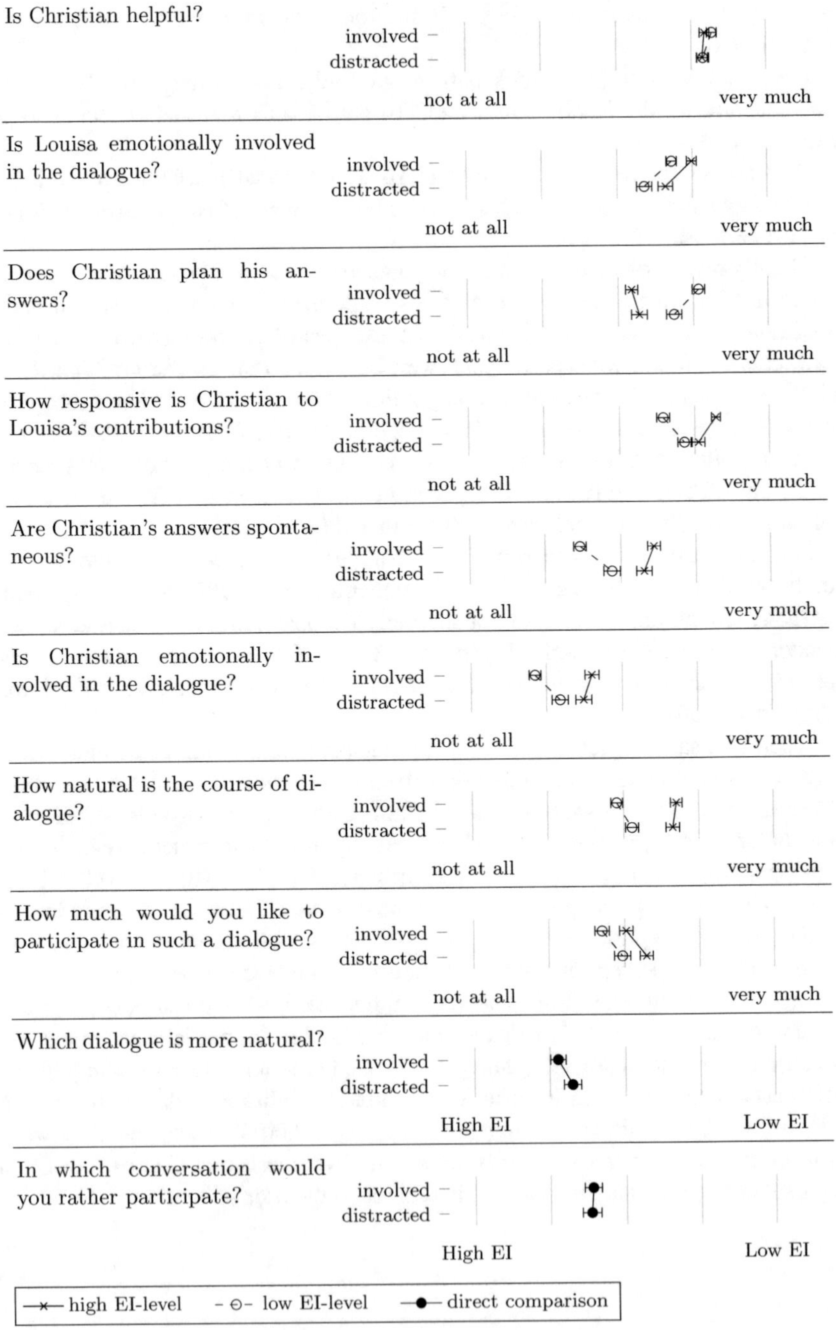

Is Christian helpful?
involved –
distracted –
not at all very much

Is Louisa emotionally involved in the dialogue?
involved –
distracted –
not at all very much

Does Christian plan his answers?
involved –
distracted –
not at all very much

How responsive is Christian to Louisa's contributions?
involved –
distracted –
not at all very much

Are Christian's answers spontaneous?
involved –
distracted –
not at all very much

Is Christian emotionally involved in the dialogue?
involved –
distracted –
not at all very much

How natural is the course of dialogue?
involved –
distracted –
not at all very much

How much would you like to participate in such a dialogue?
involved –
distracted –
not at all very much

Which dialogue is more natural?
involved –
distracted –
High EI Low EI

In which conversation would you rather participate?
involved –
distracted –
High EI Low EI

high EI-level low EI-level direct comparison

Fig. 2 Mean and standard error of the mean for the ratings of each question

2.3 Virtual Avatar Versus Human Conversation Partner

In addition to our main research questions, we compared a human and a virtual avatar as dialogue partners to check for potential differences of the impact of the EI-level in human-human and human-computer interaction.

Only Q06, 'Is Christian emotionally involved in the dialogue?', achieves a significant difference, with higher ratings for the human ($p < 0.001$). This might be due to a higher expressiveness of the human actor or to a reluctance of the participants to attribute involvement to a virtual agent.

3 Implications for Adaptive Dialogue Management

We could show in our study that the perception of a dialogue varies depending on the EI-level. This implies that the suitable EI-level should be considered in a dialogue system in order to provide a better user experience. A virtual avatar can portray different characteristics such as spontaneity or responsiveness by adjusting the EI-level.

Our study also suggests that the effects of the EI-level depend at least to some degree on the situation of the user. Although the user was still perceived as being rather involved in the *distracted* condition, a significant difference to the *involved* portrayal was reported by the participants and resulted in a significant change of the assessment of the avatar's characteristics. The perceived difference between the EI-levels is less pronounced in the *distracted* condition of our user study.

We conclude that the level of EI can indeed be utilised by an adaptive DM, as the EI level influences the user experience and does so in different ways in different situations. Exemplary, we can conceive the following behaviour for a dialogue system from our results: if the user is involved in the dialogue, a high EI-level is employed to achieve the impression that the avatar is an empathic and spontaneous conversation partner. If the user becomes distracted, the EI-level does not influence the perceived characteristics of the avatar as much. Therefore, the dialogue system can employ varying degrees of EI to keep the dialogue more diversified and interesting. Furthermore, a more pronounced change in Louisa's situation (or other conditions such as emotion) might result in different user preferences regarding the level EI that can be taken into account by adaptive dialogue management, although this assumption will have to be tested in further studies.

4 Conclusions and Future Work

Most DMs pursue a hard approach to adaptation. In this paper, we investigated whether soft approaches, such as changing the level of EI, can be successfully applied in adaptive dialogue management. We presented a user study with which we were able to ascertain the impact EI on the assessment of a dialogue and show that this assessment changes in relation to the user's situation. We conclude from our results that EI offer valuable possibilities for adaptation and therefore should be incorporated in adaptive dialogue management.

In future work, the scope of applicability of EI, e.g. as means to adapt to emotion or culture, can be assessed. Furthermore, the automatic modification of dialogue actions with the goal to produce the desired level of EI should be investigated.

Acknowledgements This paper is part of a project that has received funding from the European Union's Horizon 2020 research and innovation programme under grant agreement No 645012.

Appendix

This appendix contains the complete dialogues spoken in the videos of the user studies for both the *high EI* and the *low EI* level, in order to enable a more extensive understanding of the conducted studies.

Dialogue of the High EI Condition

SYSTEM: Hello Louisa.
USER: Hello Christian.
SYSTEM: How can I help you?
USER: I'm afraid Mr. Smith doesn't drink enough during the day.
SYSTEM: How much does Mr. Smith usually drink?
USER: Only one cup of tea after breakfast.
SYSTEM: You're right, that is not enough. Do you know why he doesn't drink enough?
USER: He doesn't like the taste of water. He would rather have some soft drink, but I don't think that is good for him.
SYSTEM: Then you should try sweetened tea. It tastes good and is not as bad as a soft drink.
USER: Perhaps I don't even need to sweeten it. He does drink tea in the morning, after all.
SYSTEM: Exactly. A little company may help as well. While sitting and chatting you can easily drink some glasses without even noticing it.
USER: I guess that's true. Chatting is just more comfortable with a nice cup of tea.
SYSTEM: By the way, does Mr. Smith use a clear glass?
USER: Yes, he does. A beer glass with a handle, so he can better grab it. Why?
SYSTEM: That might be too unobtrusive. A clear glass filled with water does not really attract attention.
USER: Do you think something more eye-catching would be better.

SYSTEM: It is worth a try. Perhaps something more colourful?
USER: I could use the cup his granddaughter gave him. It's very colourful.
SYSTEM: That is a very good idea. He will be happy whenever he sees that cup.
USER: I think so, too. He loves her very much. Also, that cup has a handle. That is very important for him, he feels more secure when grabbing something with a handle.
SYSTEM: Many elderly feel that way.
USER: But what if Mr. Smith still doesn't drink enough? After all, I need to increase that quite a lot.
SYSTEM: The body doesn't only get fluid by drinking. Food contains fluid too.
USER: I don't think that will be enough.
SYSTEM: There some dishes that contain a lot of fluid, such as soup.
USER: Or goulash? He really likes that.
SYSTEM: Yes, goulash is fine. And fruits.
USER: Alright. Thank you, Kristina. You really helped me.
SYSTEM: No problem at all.
USER: I'll get back to work.
SYSTEM: I will be here if you need me.

Dialogue of the Low EI Condition

SYSTEM: Hello Louisa.
USER: Hello Christian.
SYSTEM: How can I help you?
USER: I'm afraid Mr. Smith doesn't drink enough during the day.
SYSTEM: How much does Mr. Smith usually drink?
USER: Only one cup of tea after breakfast.
SYSTEM: You're right, that is not enough. Do you know why he doesn't drink enough?
USER: He doesn't like the taste of water. He would rather have some soft drink, but I don't think that is good for him.
SYSTEM: Then you should try sweetened tea. It tastes good and is not as bad as a soft drink.
USER: Perhaps I don't even need to sweeten it. He does drink tea in the morning, after all.
SYSTEM: Yes, you can try that.
USER: But what if Mr. Smith still doesn't drink enough? After all, I need to increase that quite a lot.
SYSTEM: I have got three more ideas. Do you want to hear all of them?
USER: Yes, please.
SYSTEM: First, you should keep him company. While chatting people usually drink a lot without even realising it.
USER: I guess that's true. Chatting is just more comfortable with a nice cup of tea.

SYSTEM: Second, you can increase his fluid intake by preparing dishes that contain much fluid, such as soup, goulash and fruits.
USER: He really likes goulash.
SYSTEM: Does Mr. Smith use a clear glass?
USER: Yes, he does. A beer glass with a handle, so he can better grab it. Why?
SYSTEM: In that case you should give him a more colourful glass. A clear glass filled with water does not really attract attention.
USER: I could use the cup his granddaughter gave him. It's very colourful.
SYSTEM: Then it is a good choice.
USER: Alright. Thank you, Kristina. You really helped me.
SYSTEM: No problem at all.
USER: I'll get back to work.
SYSTEM: I will be here if you need me.

References

1. André E, Rehm M, Minker W, Bühler D (2004) Endowing spoken language dialogue systems with emotional intelligence. In: Affective dialogue systems. Springer, pp 178–187
2. Bertrand G, Nothdurft F, Minker W, Traue H, Walter S (2011) Adapting dialogue to user emotion-a wizard-of-oz study for adaptation strategies. In: Proceedings of IWSDS, pp 285–294
3. Chu-Carroll J (2000) Mimic: an adaptive mixed initiative spoken dialogue system for information queries. In: Proceedings of the sixth conference on applied natural language processing. Association for Computational Linguistics, pp 97–104
4. Damian I, Endrass B, Huber P, Bee N, André E (2011) Individualizing agent interactions. In: Proceedings of 4th international conference on motion in games (MIG 2011)
5. Gnjatović M, Rösner D (2008) Adaptive dialogue management in the NIMITEK prototype system. In: Perception in multimodal dialogue systems. Springer, pp 14–25
6. Jaksic N, Branco P, Stephenson P, Encarnação LM (2006) The effectiveness of social agents in reducing user frustration. In: CHI'06 extended abstracts on Human factors in computing systems. ACM, pp 917–922
7. Kousidis S, Kennington C, Baumann T, Buschmeier H, Kopp S, Schlangen D (2014) A multimodal in-car dialogue system that tracks the driver's attention. In: Proceedings of the 16th international conference on multimodal interaction. ACM, pp 26–33
8. Saerbeck M, Schut T, Bartneck C, Janse MD (2010) Expressive robots in education: varying the degree of social supportive behavior of a robotic tutor. In: Proceedings of the SIGCHI conference on human factors in computing systems. ACM, pp 1613–1622
9. Turunen M, Hakulinen J (2001) Agent-based adaptive interaction and dialogue management architecture for speech applications. In: Text, speech and dialogue. Springer, pp 357–364
10. Ultes S, Minker W (2014) Managing adaptive spoken dialogue for intelligent environments. J Ambient Intell Smart Environ 6(5):523–539

An Open-Source Dialog System with Real-Time Engagement Tracking for Job Interview Training Applications

Zhou Yu, Vikram Ramanarayanan, Patrick Lange and David Suendermann-Oeft

Abstract In complex conversation tasks, people react to their interlocutor's state, such as uncertainty and engagement to improve conversation effectiveness Forbes-Riley and Litman (Adapting to student uncertainty improves tutoring dialogues, pp 33–40, 2009 [2]). If a conversational system reacts to a user's state, would that lead to a better conversation experience? To test this hypothesis, we designed and implemented a dialog system that tracks and reacts to a user's state, such as engagement, in real time. We designed and implemented a conversational job interview task based on the proposed framework. The system acts as an interviewer and reacts to user's disengagement in real-time with positive feedback strategies designed to re-engage the user in the job interview process. Experiments suggest that users speak more while interacting with the engagement-coordinated version of the system as compared to a non-coordinated version. Users also reported the former system as being more engaging and providing a better user experience.

Keywords Multimodal dialog systems · Engagement · Automated interviewing

1 Introduction and Related Work

Recently, multimodal sensing technologies such as face recognition, head tracking, etc. have improved. Those technologies are now robust enough to tolerate a fair amount of noise in the visual and acoustic background [3, 7]. So it is now possible to

Z. Yu (✉)
Carnegie Mellon University, Pittsburgh, PA, USA
e-mail: zhouyu@cs.cmu.edu; joyu@ucdavis.edu

V. Ramanarayanan · P. Lange · D. Suendermann-Oeft
Educational Testing Service (ETS) R&D, San Francisco, CA, USA
e-mail: vramanarayanan@ets.org

P. Lange
e-mail: plange@ets.org

D. Suendermann-Oeft
e-mail: suendermann-oeft@ets.org

M. Eskenazi et al. (eds.), *Advanced Social Interaction with Agents*, Lecture Notes in Electrical Engineering 510, https://doi.org/10.1007/978-3-319-92108-2_21

incorporate these technologies into spoken dialog systems to make the system aware of the user's behavior and state, which in turn will result in more natural and effective conversations [14].

Multimodal information has been proven to be useful in dialog system design in driving both low level mechanics such as turn taking as well as high level mechanics such as conversation planning. Sciutti et al. [11] used gaze as an implicit signal for turn taking in a robotic teaching context. In [17], a direction-giving robot used conversational strategies such as pause and restarts to regulate the user's attention. Kousidis et al. [4] used situated incremental speech synthesis that accommodates users' cognitive load in a in-car navigation task, which improved user experience but the task performance stays the same. In Yu et al. [19], a chatbot reacts to the user's disengagement by generating utterances that actively invite the user to continue the conversation.

Thus we propose a task-oriented dialog system framework that senses and coordinate to a user's state, such as engagement, in real time. The framework is built on top of the HALEF[1] open-source cloud-based standards-compliant multimodal dialog system framework [8, 20]. It extracts multimodal features based on data that is streamed via the user's webcam and microphone in real time. Then the system uses these multimodal features, such as gaze and spoken word count to predict a user's state, such as engagement, using a pre-built machine learning model. Then the dialog manager takes the user's state into consideration in generating the system response. For example, the system could use some conversational strategies, such as positive feedback to react to the user's disengagement state.

With the advantage of being accessible via web-browser, HALEF enables users interact with the system whenever and wherever in their comfortable environment, thus making the data collection and system evaluation process much easier and economical.

2 The HALEF Framework

In this section, we describe the sub-components of the framework. Figure 1 schematically depicts the overall architecture of the HALEF framework.

2.1 The Multimodal HALEF Framework

FreeSWITCH, specifically versions above 1.6,[2] is a scalable, open-source and cross-platform telephony framework designed to route and interconnect popular communication protocols using audio, video, text or any other form of media. FreeSWITCH allows the experimenter to modify interaction settings, such as the number of people

[1]https://sourceforge.net/projects/halef/.

[2]https://freeswitch.org/confluence/display/FREESWITCH/FreeSWITCH+1.6+Video.

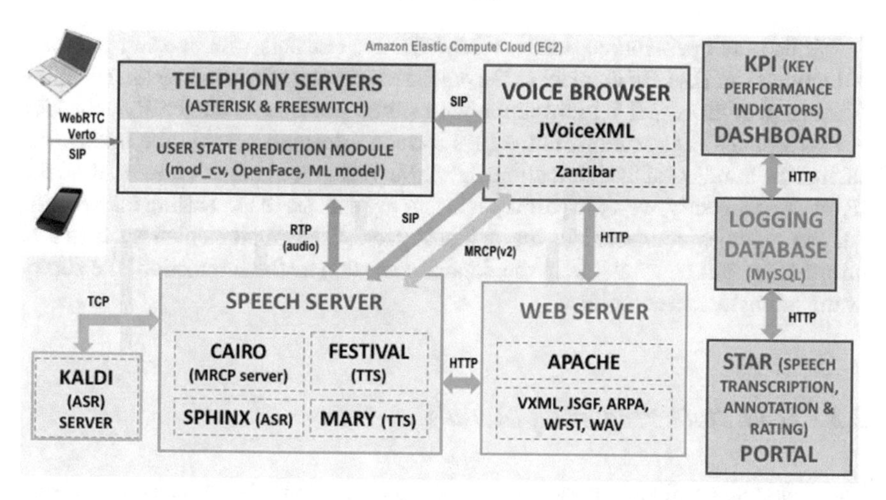

Fig. 1 System architecture of the HALEF dialog system that incorporates an engagement tracking module

who can call in at any given time, whether to display the video of the user on the webpage, the resolution of the video, sampling rate of the audio, etc. The FreeSWITCH Verto protocol also allows users to choose between different I/O devices for recording. They can switch between different microphones and cameras connected to their computers by selecting appropriate options on the web-based graphical user interface. We use FreeSWITCH Verto to connect user to HALEF via web browsers.

HALEF leverages different open-source components to form a SDS framework that is modular and industry-standard-compliant: Asterisk, a SIP (Session Initiation Protocol) and PSTN (Public Switched Telephone Network) compatible telephony server [13]; JVoiceXML, an open-source voice browser that can process SIP traffic [9]; Cairo, an MRCP (Media Resource Control Protocol) speech server, which allows the voice browser to request speech recognition, speech synthesis, audio playback and audio recording from the respective components; the Sphinx automatic speech recognizer [5] and the Kaldi ASR system; Festival [12] and Mary [10] text to speech synthesis engines; and an Apache Tomcat-based web server that can host dynamic VoiceXML (VXML) pages and serve media files such as grammars and audio files to the voice browser. OpenVXML allows designers to author the dialog workflow as a flowchart, including details of specific grammar files to be used by the speech recognizer and text-to-speech prompts that need to be synthesized. In addition, dialog designers can insert "script" blocks of Javascript code into the workflow that can be used to perform simple processing steps, such as creating HTTP requests to make use of natural language understanding web services on speech recognition output. In order to react to the user's engagement, these "script" blocks retrieve and act upon the engagement score of the user in real time. The entire workflow can be exported to a Web Archive (or WAR) application, which can then be deployed on an Apache Tomcat web server that serves Voice XML documents.

We use the open-source database MySQL for our data warehousing purposes. All modules in HALEF connect to the database and write their log messages into it. We then post-process this information with stored procedures into easily accessible views. Metadata extracted from the logs include information about the interactions such as the duration of the interaction, the audio/video recording file names, the caller IP, etc. Additionally, we store participants' survey data and expert rating information. All the modules connected to the database have been implemented such that all information will be available in the database as soon as the interaction, the survey, or the rating task is completed.

2.2 User State Prediction Module

The user state prediction module is linked with FreeSWITCH via sockets to a standalone Linux server for automatic head tracking using OpenFace [1]. The head tracker receives raw images from FreeSWITCH and performs tracking of the user's head movement, gaze direction and facial action units. Visual information has been shown to be critical in assessing the mental states of the users in other systems as well [18]. So we include visual information in predicting the user state. The system uses the pre-trained machine learning model to predict the user's state. The user state module doesn't connect to HALEF directly, it passed the engagement information to the database first and then the application retrieves the user state information from the database. The application selects the conversation branch based on the user state information as well as the spoken language understanding results.

3 Example Application: Job Interview Training/Practice

In this conversation task, the system acts as the interviewer for a job in a pizza restaurant. The system first asks the user some basic personal information and then proposes two scenarios about conflicts that may happen in the workplace and asks the user how he/she would resolve them. We designed the task to assess non-native speakers' English conversational skills, pragmatic appropriateness of responses, and their ability to comprehend the stimulus materials and respond appropriately to questions posed during naturalistic conversational settings.

3.1 User Engagement Modeling

We first collected a set of data using a non-coordinated multimodal interviewer version. We then used this dataset to build supervised machine learning models to predict engagement in real time, as well as a baseline for the reactive version of the system. We collected 200 video recordings in all from crowdsourced participants recruited via the Amazon Mechanical Turk platform. To train the engagement model we randomly selected 30 conversations which satisfied the following two quality criteria: (i) the face recognition system detects a face with high confidence (80%) throughout the interaction, and (ii) the automatic speech recognition output of the user utterances is not empty. There are in total 367 conversational exchanges in total over 30 conversations (note that we use the term conversational exchange here to denote a pair of one system turn and one user turn). We asked three experts to annotate user engagement for every conversational exchange based on the video and audio recordings. We adopted the engagement definition and annotation scheme introduced in [19]. For the purposes of our study, we defined engagement as the degree to which users are willing to participate in the task along three dimensions—behaviorally (staying on task and following directions), emotionally (for instance, not being bored by the task) and cognitively (maximizing their cognitive abilities, including focused attention, memory, and creative thinking) [16]. Ratings were assigned on a 1–5 Likert scale ranging from very disengaged to very engaged. We collapsed 1–2 ratings into a "disengaged" label, and 3–5 into an "engaged" label, because the system is designed to react to a binary signal in the conversational flow. The threshold was chosen because we would like to only regulate the extreme cases in our task, in order to keep the conversation to be effective. For other tasks, we recommend setting the threshold as an experimental parameter that decided through user preference. There are in total three annotators involved and they had an inter-annotator Cohen's κ agreement value of 0.82 on average on the binary engaged vs. disengaged rating task. For modeling purposes, we used the average label from all annotators as the ground truth or gold standard label. Among all the conversation exchanges, 75% were labeled as "engaged" and 25% were labeled as "disengaged". So while the current work only considers binary labels, future work will examine design policies that takes a finer grained 5-point engagement scale into consideration.

In order to train a vision-based engagement predictor, we extracted the following vision features: head pose, gaze and facial action units. After the user state module receives the real-time features, it simultaneously performs three computing steps for engagement detection [1]. It processes the mean and variance of the head pose change to determine the frequency of users changing their head pose. It also extracts the mean and variance of the action units that relate to smiles, in order to calculate the frequency of user smiles. It further computes the mean and variance of the gaze direction, to capture how frequency users shift their gaze. These features are computed per conversational exchange to form the feature set for engagement predictor training. Further, we also take verbal information into account for engagement prediction. In this interview training task, since there are a fixed number of conversation states, we calculated the mean word count of all the conversations that are labeled as disengaged in each state. We use this value as the threshold to decide if the user is disengaged or not for each state. The verbal engagement score is computed over the ASR output as soon as the user utterance is finished.

We used leave-one-conversation-out cross validation and a Support Vector Machine with linear kernel to train the model. Our experiments resulted in an F1 measure of 0.89 (the majority vote baseline was 0.72). We observed that the failures occurred mainly due to the system turn-taking errors, such as system interrupting the user, which results a shorter user response and in turn leads to a low engagement score in the verbal channel. Note that the relative high baseline is due to the skewness of the data, as there are more engaged conversational exchanges than disengaged ones. During testing, we quantify vision features in a time window which is empirically determined by the engagement predictor's performance based on the task (we chose 2 s, which happens to be the mean of the conversational exchange duration). Then we combine all the multimodal features mentioned above with weights obtained from the machine learning model to obtain a score that represents the visual engagement of the user. The dialog manager receives a score which is a weighted sum of all the modality-wise engagement scores with a set of weights determined empirically.

As a follow-up experiment to see how well this initial engagement detector performed, we then collected 54 conversations using the engagement-coordinated interview training application. Among them we found 23 recordings that satisfied the two quality criteria mentioned earlier as well as an additional criterion of analyzing data from unique speakers. Experts then rated the engagement at each conversational exchange where the system is required to make a dialog management decision based on the user's predicted engagement state.

Next, we used this data to retrain the linear regression weights assigned for the vision and verbal modalities in Eq. (1); here x_{i1} stands for the visual-engagement value, x_{i2} stands for the verbal-engagement value and y_i stands for the ground truth label. The simple linear regression analysis performed a least-squares optimization of the following cost function:

$$min_{\alpha\beta} \sum_{i=1}^{n} (y_i - \alpha x_{i1} - \beta x_{i2})^2 \tag{1}$$

We found optimal regression coefficients of $\alpha = 0.63$ and $\beta = 0.37$ and adjusted the weights in the model accordingly. We then collected another set of 50 conversations with adjusted weights. Among them 32 of them are valid videos for analysis according to the quality criteria. In the result analysis we used this batch data for analysis for the engagement-coordinated system. We found the F1 score for the trained engagement classifier was 0.86, a significant improvement over the majority vote baseline method of 0.74.

3.2 *Conversational Strategy Design*

The communication and education literature mention a number of conversational strategies which are useful to improve user engagement, such as active participation, mention of shared experience [15], and positive feedback and encouragement [6], among others. Particularly in job interview literature, researchers find that infusing positive sentiments into the conversation could lead to more self-disclosure from interviewees. With this in mind, we designed a set of positive feedback strategies with respect to different dialog states. For example, in one dialog state, we asked about the interviewee's previous experience.

3.3 Coordination Policy

We implemented a local greedy policy to react to the user's disengagement in this interview training task. Once the dialog manager receives the signal from the end-pointing module reporting that the user has finished the turn, it queries the most recent engagement score from the database. If the user is sensed as disengaged, the positive strategy that is designed with respect to that conversational state will trigger, otherwise the conversation goes into next dialog state.

3.4 Results

We asked callers to fill out a survey after interacting with the system. We asked them how engaged they felt overall during the interaction as well as their overall conversational experience on a 1–5 Likert scale. We compared the user responses of the 30 conversations that were collected using the non-coordinated interview training method to the 32 conversations that are collected using the engagement-coordinated version, and found that the engagement-coordinated system received statistically higher overall ratings from the users in terms of both overall engagement and user experience (see Fig. 2 for details).

Though the engagement-coordinated version had more system utterances than the non-coordinated one, the extra utterances are all statements (e.g., "I think the manager would do that.") instead of questions, so no extra user utterances were elicited. Nonetheless, when we calculated the number of unique tokens of all the uses' utterances based on the ASR output, we found that users who interacted with the engagement-coordinated version expressed significantly more information than users who interacted with the non-coordinated version (see Fig. 2 for details).

We also found that there were three users who interacted with the engagement-coordinated interview item more than once. We found that their assessed average engagement improved (from 2.3 to 4.0) after interacting with the system several times. This gives us a positive indication that interacting with our system does have the potential to allow users to improve their conversational ability during job interviews.

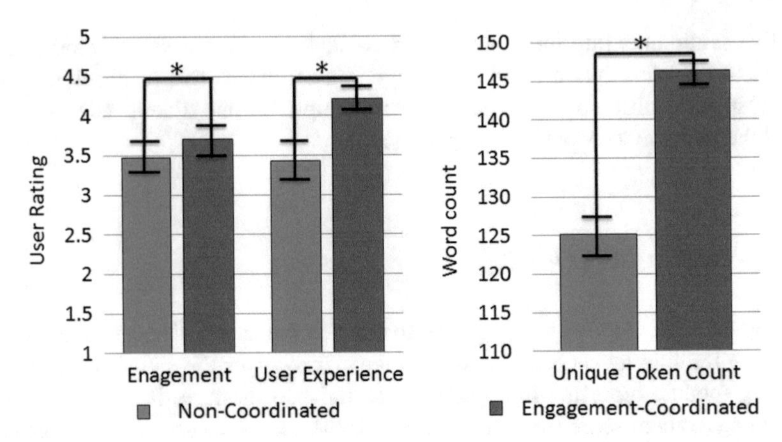

Fig. 2 Experiment results between non-coordinated and engagement-coordinated job interviewer

4 Conclusion and Future Work

We have proposed and implemented a real-time user-reactive system framework for task-oriented conversations. We implemented an example application based on the framework in the form of an engagement-coordinated interview training task. From the data collected using both the non-coordinated and engagement-coordinated versions of the interview training task, we found that the engagement-coordinated version was rated as more engaging and providing a better user experience as compared to a non-coordinated system.

In the future, we wish to design and implement improved reactive systems that are able to tackle more complex tasks, such as for conversational proficiency practice. We also wish to design better conversation policies that take dialog context information into consideration during conversation flow planning. We will also integrate more speech feature-based information into the engagement module in order to make the engagement prediction more accurate.

Acknowledgements We would like to thank Robert Mundkowsky and Dmytro Galochkin for help with system engineering. We would also like to thank Eugene Tsuprun, Keelan Evanini, Nehal Sadek and Liz Bredlau for help with the task design and useful discussions.

References

1. Baltru T, Robinson P, Morency L-P et al (2016) Openface: an open source facial behavior analysis toolkit. In: 2016 IEEE winter conference on applications of computer vision (WACV). IEEE, pp 1–10
2. Forbes-Riley K, Litman DJ (2009) Adapting to student uncertainty improves tutoring dialogues. In: AIED, pp 33–40

3. He X, Yan S, Hu Y, Niyogi P, Zhang H-J (2005) Face recognition using laplacian faces. IEEE Trans Pattern Anal Mach Intell 27(3):328–340
4. Kousidis S, Kennington C, Baumann T, Buschmeier H, Kopp S, Schlangen D (2014) A multimodal in-car dialogue system that tracks the driver's attention. In: Proceedings of the 16th international conference on multimodal interaction. ACM, pp 26–33
5. Lamere P, Kwok P, Gouvea E, Raj B, Singh R, Walker W, Warmuth M, Wolf P (2003) The cmu sphinx-4 speech recognition system. In: IEEE international conference on acoustics, speech and signal processing (ICASSP 2003), Hong Kong, vol 1, pp 2–5 (Citeseer)
6. Lehman B, DMello S, Graesser A (2012) Interventions to regulate confusion during learning. In: International conference on intelligent tutoring systems. Springer, pp 576–578
7. Morency L-P, Sidner C, Lee C, Darrell T (2005) Contextual recognition of head gestures. In: Proceedings of the 7th international conference on multimodal interfaces. ACM, pp 18–24
8. Ramanarayanan V, Suendermann-Oeft D, Lange P, Mundkowsky R, Ivanou A, Yu Z, Qian Y, Evanini K (2016) Assembling the jigsaw: how multiple w3c standards are synergistically combined in the halef multimodal dialog system. In: Multimodal interaction with W3C standards: towards natural user interfaces to everything. Springer, page to appear
9. Schnelle-Walka D, Radomski S, Mühlhäuser M (2013) Jvoicexml as a modality component in the w3c multimodal architecture. J Multimodal User Interfaces 7(3):183–194
10. Schröder M, Trouvain J (2003) The german text-to-speech synthesis system mary: a tool for research, development and teaching. Int J Speech Technol 6(4):365–377
11. Sciutti A, Schillingmann L, Palinko O, Nagai Y, Sandini G (2015) A gaze-contingent dictating robot to study turn-taking. In: Proceedings of the tenth annual acm/ieee international conference on human-robot interaction extended abstracts. ACM, pp 137–138
12. Taylor P, Black AW, Caley R (1998) The architecture of the festival speech synthesis system
13. Van Meggelen J, Madsen L, Smith J (2007) Asterisk: the future of telephony. O'Reilly Media, Inc
14. Vinciarelli A, Pantic M, Bourlard H (2009) Social signal processing: survey of an emerging domain. Image Vis Comput J 27(12):1743–1759
15. Wendler D (2014) Improve your social skills. CreateSpace Independent Publishing Platform
16. Whitehill J, Serpell Z, Lin Y-C, Foster A, Movellan JR (2014) The faces of engagement: automatic recognition of student engagementfrom facial expressions. IEEE Trans Affect Comput 5(1):86–98
17. Yu Z, Bohus D, Horvitz E (2015) Incremental coordination: attention-centric speech production in a physically situated conversational agent. In: 16th annual meeting of the special interest group on discourse and dialogue, p 402
18. Yu Z, Gerritsen D, Ogan A, Black AW, Cassell J (2013) Automatic prediction of friendship via multi-model dyadic features. In: Proceedings of SIGDIAL, pp 51–60
19. Yu Z, Nicolich-Henkin L, Black A, Rudnicky A (2016) A wizard-of-oz study on a non-task-oriented dialog systems that reacts to user engagement. In: 17th annual meeting of the special interest group on discourse and dialogue
20. Yu Z, Ramanarayanan V, Mundkowsky R, Lange P, Ivanov A, Black AW, Suendermann-Oeft D (2016) Multimodal halef: an open-source modular web-based multimodal dialog framework

Part V
Advanced Dialogue System Architecture

On the Applicability of a User Satisfaction-Based Reward for Dialogue Policy Learning

Stefan Ultes, Juliana Miehle and Wolfgang Minker

Abstract Finding a good dialogue policy using reinforcement learning usually relies on objective criteria for modelling the reward signal, e.g., task success. In this contribution, we propose to use user satisfaction instead represented by the metric Interaction Quality (IQ). Comparing the user satisfaction-based reward to the baseline of task success, we show that IQ is a real alternative for reward modelling: designing a reward function using IQ may result in a similar or even better performance than using task success. This is demonstrated in a user simulator evaluation using a live IQ estimation module.

Keywords Statistical spoken dialogue systems · Interaction quality · Reinforcement learning

1 Introduction and Related Work

Finding well-performing dialogue strategies for Spoken Dialogue Systems (SDSs) has been in the focus of research for many years. One possibility is to create handcrafted rules designed by experts. However, this approach is problematic: the resulting strategy is usually not very robust and strongly biased by the expert's view.

Instead of handcrafting the dialogue behaviour, more recent approaches aim at learning the optimal dialogue behaviour using reinforcement learning [6, 24]. Here, the dialogue strategy (called policy) is trained with a number of sample dialogues which are evaluated using a reward R. Traditional approaches incorporate the objective task success (TS) into the reward. This task success, though, only models the system view on the interaction. Incorporating the user view instead might be of more

S. Ultes (✉)
Department of Engineering, University of Cambridge, Trumpington Street, Cambridge, UK
e-mail: su259@cam.ac.uk

J. Miehle · W. Minker
Institute of Communications Engineering, Ulm University, Albert-Einstein-Allee 43, Ulm, Germany

M. Eskenazi et al. (eds.), *Advanced Social Interaction with Agents*, Lecture Notes in Electrical Engineering 510, https://doi.org/10.1007/978-3-319-92108-2_22

interest [15, 19] as it is the user who ultimately decides whether to use the system again or not.

For task-oriented dialogue systems, the user view may be captured with the user satisfaction. In fact, task success has only been used in previous work as it has been found to correlate well with user satisfaction [23]. While previously, user satisfaction was not accessible during learning, the recently proposed Interaction Quality (IQ) metric [10] has been designed to automatically predict the user's satisfaction level during the dialogue.

The main contribution of this work is to address the question how user satisfaction and task success relate with respect to their applicability on dialogue policy learning when being used as the main reward. We formulate the following expectations on their behaviour:

1. A policy which is optimised on user satisfaction should yield similar performance in task success compared to a policy optimised on task success. We assume that a user is not satisfied with the dialogue if the dialogue was not successful.
2. A policy which is optimised on task success will result in worse satisfaction rates than a policy which is optimised on user satisfaction.

Hence, for a satisfaction metric to be applicable for dialogue learning, it must show similar behaviour. In our previous work, we successfully showed that IQ is suitable for designing a rule-based policy and outlined an IQ-based dialogue-level reward function [16] and showed that using IQ for a binary decision over TS results in a precision of 84.5%. This indicates that similar performance in task success rate may be expected when using it as the main reward.

To follow up on this, we present a study in a simulated environment comparing IQ and TS as the main rewards for dialogue learning. We investigate the effects on the resulting TS rate of using IQ as the main reward and vice versa. Only if the expected behaviour is met, applying IQ for dialogue learning may be regarded as feasible.

Others have previously introduced user ratings into the reward. Gašić et al. [4] have successfully used the user's success rating directly during learning with real humans. Su et al. [12] extended the idea by using a similar setup plus having an additional task success estimator. While both use task success as measure, we will investigate whether user satisfaction may be used as reward in a similar fashion.

A prominent way to model user satisfaction is the PARADISE framework [22] which has also been used for reward modelling [8, 21, e.g.]. However, a questionnaire has to be answered by real users to derive user ratings with that framework. This is usually not feasible in real world setting. To overcome this, PARADISE has been used in conjunction with expert annotators [2, 3] which allows an unintrusive rating. However, the problem of mapping the results of the questionnaire to a scalar reward value still exists. Furthermore, PARADISE assumes a linear dependency between measurable parameters and user satisfaction. However, assuming a non-linear dependency might be more appropriate [9]. Finally, to derive a user satisfaction rating using the PARADISE framework requires access to information about the success of the dialogue which us usually hard to obtain. Therefore, we will use the Interaction Quality [10] in this work which uses scalar values applied by experts and assumes a non-linear dependency between measurable parameters and the target value and does not rely on any information about the success of the dialogue or any task specific information.

The remainder of the paper is organized as follows: the IQ metric and the core contribution of modelling dialogue-level reward functions with IQ is described in Sect. 2 with its experiments and results in Sect. 3. Conclusions for future work are drawn in Sect. 4.

2 Interaction Quality Dialogue-Level Reward

The Interaction Quality (IQ) [10], which will be used for modelling the dialogue-level reward, has been proposed as a turn-level metric for user satisfaction in spoken dialogue systems.

Interaction Quality Paradigm The general idea of the IQ paradigm is to use a set of measurable interaction parameters to create a non-linear statistical classification model. The target variable is a scalar value ranging from five (=satisfied) down to one (=completely unsatisfied). The input variables called interaction parameters encode information about the current turn as well as temporal information which is modelled on the window and the dialogue level (counts, means, sums and rates of turn-level parameters). The turn-level parameters are derived from the SDS modules Speech Recognition, Language Understanding, and Dialogue Management.

IQ meets the requirements for being used in an adaptive dialogue framework [19] and is the ideal candidate for our research. The IQ values are annotated by expert annotators yielding a high correlation to real user ratings [20].

Previously, a Support Vector Machine (SVM) has been applied for IQ classification achieving an unweighted average recall[1] (UAR) of 0.59 [9]. Furthermore, an ordinal regression approach achieved 0.55 UAR [1] and a hybrid-HMM 0.51 UAR [17]. As comparison, the human performance on that task is 0.69 UAR [10].

Reward Model In this work, we compare two different approaches for modelling the reward. Both have the same shape: for each turn, a small negative discount is added to favour shorter dialogues. Additionally, a final reward is defined based on the dialogue outcome. Both are combined for calculating the overall reward R for a complete dialogue of length T. For the baseline of using task success (TS), R is defined as

$$R_{TS} = T \cdot (-1) + \mathbb{1}_{TS} \cdot 20 \,, \tag{1}$$

where $\mathbb{1}_{TS} = 1$ only if the dialogue resulted in a successful task, $\mathbb{1}_{TS} = 0$ otherwise.

Based on the same binary decision principle, the IQ-based reward function has been defined:

[1]The arithmetic average over all class-wise recalls.

$$R_{IQ_b} = T \cdot (-1) + \mathbb{1}_{IQ} \cdot 20 \,. \tag{2}$$

A final reward of 20 is assigned only if a high IQ has been achieved at the end of a dialogue, i.e., the final IQ value ($\mathbb{1}_{IQ} = 1$ only if $IQ \geq 4$, otherwise $\mathbb{1}_{IQ} = 0$).

3 Experiments and Results

Evaluation of the reward functions presented in Sect. 2 is performed using an adaptive dialogue system interacting with a user simulator. A user simulator offers an easy and cost-effective way for training and evaluating dialogue policies and getting a basic impression about their performance.

Experimental Setup For creating the IQ estimation model as well as for training and evaluation of the dialogues, the Let's Go bus information domain [7] has been chosen as it represents a domain of suitable complexity. The Let's Go User Simulator (LGUS) [5] is used for policy training and evaluation to neutralise the need for human evaluators. LGUS has been trained on 1,275 real user dialogues collected with Let's Go. The simulator is set to converse for at least 5 turns as this is the minimum number to successfully complete the dialogue. To get bus information from the system, departure place, arrival place, travel time, and optionally the bus number may be provided.

In order to evaluate the reward functions, the adaptive statistical dialogue manager OwlSpeak [18] is used with a connected IQ estimation module [14]. The IQ estimation module uses a Support Vector Machine UAR of 0.55 on the training data [11] using 10-fold cross-validation. The policy of OwlSpeak operates on the summary level, i.e., it maps a summary space representation to a summary action, e.g., `request` or `confirm`. The summary action is mapped back to a system action using a heuristic.

For evaluation, the objective metrics task success rate (TSR, the ratio of dialogues for which the system was able to provide the correct result) and average dialogue length (ADL) have been chosen. In addition, the average IQ value (AIQ) is calculated for each policy based on the IQ value of the last exchange of each dialogue (which is also used within some of the reward functions). All AIQ values are based on the SVM estimates.

Results For each reward model, the results depicted in Table 1 are computed after 1,000 training dialogues based on the last 200 dialogues averaged over five trails. The corresponding learning curves for moving TSR and IQ are shown in Fig. 1. Both show that for the respective reward, learning has mostly saturated for the last 200 dialogues.

Looking at the relation between IQ and TSR for the two reward models, both expectations presented in Sect. 1 are clearly met: R_{IQ} results in similar success rates compared to R_{TS} (Exp. 1) and yields higher IQ values than R_{IQ} (Exp. 2) with 2.0 compared to 1.5. Even though the success rate for R_{IQ} with an TSR of 56.3% sur-

Table 1 Final TSR, ADL, and AIQ computed out of the last 200 training dialogues for each reward function averaged over five policy trainings along with the respective confidence interval

	TSR	ADL	AIQ
IQ_b	56.3% (±11.8)	13.3 (±0.9)	2.0 (±0.5)
TS	45.4% (±9.5)	14.2 (±1.4)	1.5 (±0.5)

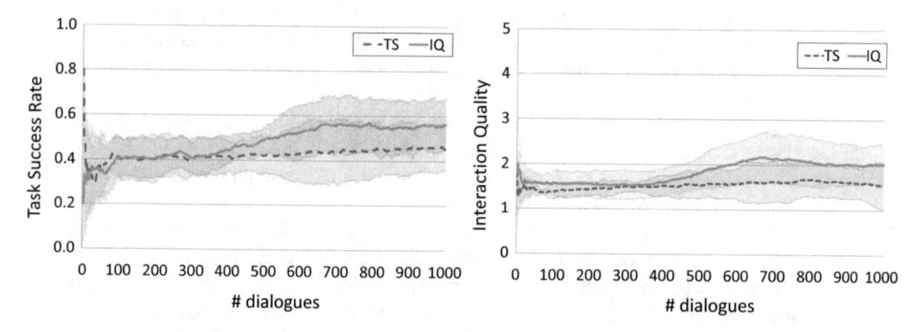

Fig. 1 Moving task success rate and moving interaction quality computed over a window of 200 turns for both reward models. The learning curves are averaged over five trained policies

passes R_{TS} with a TSR of 45.4%, this may regarded as a statistical insignificant. In fact, using the same trained reward estimator to learn policies in other domains shows that the TSR of R_{IQ} and R_{TS} are similar [13].

As both expectations have been met, the results clearly suggest that IQ is applicable for dialogue policy learning. Although the experiments have only been carried out in simulation, we would expect to see similar behaviour in real user experiments.

4 Conclusion

In this paper, we have presented user satisfaction represented by the Interaction Quality (IQ) metric as an alternative to task success for modelling the dialogue-level reward used for reinforcement learning of the dialogue policy. We have analysed its applicability by formulating two key expectations on the relation of task success and IQ and shown in simulated experiments that these expectations have been clearly met. Learning a dialogue policy using IQ in the reward results in similar performance of the resulting policy in terms of task success while achieving better results in terms of estimated user satisfaction.

Of course, to asses the impact on user satisfaction, experiments with real users are necessary which will be part of future work. Furthermore, while the difference in TSR is not significant, it needs to be investigated further. Finally, IQ is currently only regarded as being part of the reward. It may as well be part of the dialogue state which will also be in the focus of future work.

References

1. El Asri L, Khouzaimi H, Laroche R, Pietquin O (2014) Ordinal regression for interaction quality prediction. In: Proceedings of ICASSP. IEEE, pp 3245–3249
2. El Asri L, Laroche R, Pietquin O (2012) Reward function learning for dialogue management. In: Proceedings of the 6th STAIRS. IOS Press, pp 95–106
3. El Asri L, Laroche R, Pietquin O (2013) Reward shaping for statistical optimisation of dialogue management. In: Statistical language and speech processing. Springer, pp 93–101
4. Gašić M, Breslin C, Henderson M, Kim D, Szummer M, Thomson B, Tsiakoulis P, Young SJ (2013) On-line policy optimisation of Bayesian spoken dialogue systems via human interaction. In: Proceedings of ICASSP. IEEE, pp 8367–8371
5. Lee S, Eskenazi M (2012) An unsupervised approach to user simulation: toward self-improving dialog systems. In: Proceedings of 13th SIGDial. ACL, pp 50–59
6. Levin E, Pieraccini R (1997) A stochastic model of computer-human interaction for learning dialogue strategies. Eurospeech 97:1883–1886
7. Raux A, Bohus D, Langner B, Black AW, Eskenazi M (2006) Doing research on a deployed spoken dialogue system: one year of let's go! experience. In: Proceedings of ICSLP
8. Rieser V, Lemon O (2008) Learning effective multimodal dialogue strategies from wizard-of-oz data: bootstrapping and evaluation. In: Proceedings of 46th ACL-HLT. ACL, pp 638–646
9. Schmitt A, Schatz B, Minker W (2011) Modeling and predicting quality in spoken human-computer interaction. In: Proceedings of 12th SIGDial. ACL, Portland, OR, pp 173–184
10. Schmitt A, Ultes S (2015) Interaction quality: assessing the quality of ongoing spoken dialog interaction by experts–and how it relates to user satisfaction. Speech Commun 74:12–36. https://doi.org/10.1016/j.specom.2015.06.003
11. Schmitt A, Ultes S, Minker W (2012) A parameterized and annotated spoken dialog corpus of the cmu let's go bus information system. In: Proceedings of LREC, pp 3369–337
12. Su PH, Gašić M, Mrkšić N, Rojas-Barahona L, Ultes S, Vandyke D, Wen TH, Young S (2016) On-line active reward learning for policy optimisation in spoken dialogue systems. In: Proceedings of 54th ACL. ACL, pp 2431–2441
13. Ultes S, Budzianowski P, Casanueva I, Mrkšić N, Rojas-Barahona L, Su PH, Wen TH, Gašić M, Young S (2017) Domain-independent user satisfaction reward estimation for dialogue policy learning. In: Proceedings of Interspeech. ISCA, pp 1721–1725. https://doi.org/10.21437/Interspeech.2017-1032
14. Ultes S, Dikme H, Minker W (2016) Dialogue management for user-centered adaptive dialogue. In: Situated dialog in speech-based human-computer interaction. Springer International Publishing, Cham, pp 51–61. https://doi.org/10.1007/978-3-319-21834-2_5
15. Ultes S, Heinroth T, Schmitt A, Minker W (2011) A theoretical framework for a user-centered spoken dialog manager. In: Proceedings of the paralinguistic information and its integration in spoken dialogue systems workshop. Springer New York, pp 241–246. https://doi.org/10.1007/978-1-4614-1335-6_24
16. Ultes S, Kraus M, Schmitt A, Minker W (2015) Quality-adaptive spoken dialogue initiative selection and implications on reward modelling. In: Proceedings of 16th SIGDIAL. ACL, pp 374–383
17. Ultes S, Minker W (2014) Interaction quality estimation in spoken dialogue systems using hybrid-HMMs. In: Proceedings of 15th SIGDIAL. ACL, pp 208–217
18. Ultes S, Minker W (2014) Managing adaptive spoken dialogue for intelligent environments. J Ambient Intell Smart Environ 6(5):523–539. https://doi.org/10.3233/ais-140275
19. Ultes S, Schmitt A, Minker W (2012) Towards quality-adaptive spoken dialogue management. In: NAACL-HLT workshop on future directions and needs in the spoken dialog community: tools and data (SDCTD 2012). ACL, Montréal, Canada, pp 49–52
20. Ultes S, Schmitt A, Minker W (2013) On quality ratings for spoken dialogue systems—experts versus users. In: Proceedings of the 2013 NAACL-HLT. ACL, pp 569–578
21. Walker M (2000) An application of reinforcement learning to dialogue strategy selection in a spoken dialogue system for email. J Artif Intell Res 12:387–416

22. Walker M, Litman DJ, Kamm CA, Abella A (1997) PARADISE: a framework for evaluating spoken dialogue agents. In: Proceedings of 8th EACL. ACL, Morristown, NJ, USA, pp 271–280. https://doi.org/10.3115/979617.979652
23. Williams JD, Young SJ (2004) Characterizing task-oriented dialog using a simulated asr chanel. In: Proceedings of 8th Interspeech, pp 185–188
24. Young SJ, Gašić M, Thomson B, Williams JD (2013) POMDP-based statistical spoken dialog systems: a review. Proc IEEE 101(5):1160–1179

OntoVPA—An Ontology-Based Dialogue Management System for Virtual Personal Assistants

Michael Wessel, Girish Acharya, James Carpenter and Min Yin

Abstract *Dialogue management (DM) is a difficult problem.* We present *OntoVPA*, an Ontology-Based *Dialogue Management System (DMS)* for *Virtual Personal Assistants (VPAs)*. The features of OntoVPA are offered as generic solutions to core DM problems, such as *dialogue state tracking, anaphora and coreference resolution,* etc. To the best of our knowledge, OntoVPA is the first commercially available, fully implemented DMS that employs ontologies and ontology-based rules for (a) domain model representation and reasoning, (b) dialogue representation and state tracking, and (c) response generation. OntoVPA is a *declarative, knowledge-based system* which can be customized to a new VPA domain by modifying and exchanging ontologies and rule bases, with *very little to no conventional programming required*.

Keywords Dialogue management systems · Virtual Personal Assistants Ontology · Knowledge representation and reasoning for dialogue management SPARQL · Web Ontology Language · OWL · Description logics

1 Introduction and Motivation

Dialogue management (DM), the core functionality of a *Dialogue Management System (DMS)*, is notoriously difficult, if not AI-complete; see [1] for a recent overview. Even in more restricted dialogue systems, difficult problems such as *anaphora and coreference resolution* and *dialogue state tracking* may have to be addressed. The importance and difficulty of the dialogue state tracking problem is also testified by

---

M. Wessel (✉) · G. Acharya · J. Carpenter · M. Yin
SRI International, 333 Ravenswood Avenue, Menlo Park, CA 94025, USA
e-mail: michael.wessel@sri.com

G. Acharya
e-mail: girish.acharya@sri.com

J. Carpenter
e-mail: james.carpenter@sri.com

M. Yin
e-mail: min.yin@sri.com

M. Eskenazi et al. (eds.), *Advanced Social Interaction with Agents*, Lecture Notes in Electrical Engineering 510, https://doi.org/10.1007/978-3-319-92108-2_23

the recently established series of *Dialog State Tracking Challenges* [2], aimed at catalyzing progress in this area.

To the best of our knowledge, state tracking and DM are still in its infancy in contemporary *commercial* frameworks/platforms for realizing *Virtual Personal Assistants (VPAs),* with very little to no support offered by the frameworks. Commercial platforms usually offer some form of action-reaction trigger (≈production) rules (e.g., to invoke a RESTful service in the Internet of Things), based on the user's current input or intent. However, these rules are usually constrained to accessing the parameters (≈slot values) of the current user intent and/or input only, and hence cannot reason over or introspect the full dialogue history or domain model. Their *context* is limited to the current (and maybe the previous) utterance/dialog step. In contrast, the *ontology-based DMS* that we are presenting in this paper, *OntoVPA,* has access to and is capable of reasoning over the full dialogue history of utterances/dialogue steps, and corresponding domain models as well, hence enabling superior *dialogue understanding and conversational capabilities* as required for *conversational VPAs*.

Consequently, in these simpler commercial systems, the VPA application developer must manually program most of the DM "bookkeeping" (i.e., dialogue state management and tracking), with little to no support from the underlying VPA framework. In the worst case, the dialogue history and state needs to be encoded in proprietary data structures, with little to no reuse across domains.

This *model-less, programmatic approach to DM* is feasible for simple *one-shot request/response* VPAs that do not need to sustain complex conversations (possibly involving multiple dialogue steps) with the user for fulfilling a request, but becomes tedious (if not impractical) in more complex VPA domains that require elaborate workflows for solving complex domain-specific problems in cooperation with the user (e.g., booking of a business trip includes flight, hotel, and rental car reservations, etc.) A system with deep domain knowledge capable of solving complex domain-specific problems in cooperation with the user is also called a *Virtual Personal Specialist (VPS)* [3]. Due to the complexity of the domain problems to be solved (e.g., bank transfers), dialogues will require multiple steps for workflow/intent completion, and both the user and the VPA should be able to steer and drive the dialogue to solve the problem cooperatively, i.e., *mixed initiative dialogue systems* [4] are preferred. The challenges presented by conversational VPSs have shaped *OntoVPA, our ontology-based DMS for VPAs.*

OntoVPA is offered as a generic, reusable VPA platform (architecture, framework, toolkit, …) for implementing conversational VPAs and VPSs that require deep domain knowledge, complex workflows, and flexible—not hard-coded—DM strategies. It promises to significantly reduce development times, due to its reusable, powerful, and generic features (see Sect. 2).

OntoVPA employs ontologies for *dialogue and domain representation,* as well as *ontology-based rules* for *dialogue management, state tracking, and response/answer computation.* OntoVPA's dialogue representation is similar to the *blackboard* in *information state space-based DMSs*, but also shares many characteristics of *frame-based DMSs* [1]. The representational and inferential capabilities of OntoVPA greatly exceed typical *frame-based DMSs* though, as its architecture includes a proper ontology representation and reasoning infrastructure. In a sense, OntoVPA takes the frame-based DMS architecture to the extreme.

Related Work Since the days of the LUNAR system, ontologies have been used successfully in NL Question Answering systems, recently in HALO/AURA [5]. One of the few dialogue systems that uses ontologies at runtime for response generation is [6], but neither standard reasoning nor ontology-based queries/rules are used. Dynamic use of ontologies is considered in [7] and led to the OntoChef system [8], which is equipped with a sophisticated cooking ontology, but uses a DMS without ontology (Olympus/RavenClaw). Other case studies have focused on ontology modeling [9] and dialogue design based on task structures in OWL [10]. Frequently, ontologies are used for domain models. OWLSpeak [11, 12] is similar to our dialogue ontology in that it is capable to represent (speech act influenced) dialogue models as well as the state of the dialogue, but no ontology-based rules nor queries are used to compute the system's responses (it generates Voice-XML). A sophisticated Semantic Dialogue System for Radiologists is presented in [13], and it also relies on ontologies and SPARQL, but does not use these techniques for the implementation of the DM shell.

The remainder of this paper is structured as follows. First, we introduce ontology key concepts and illustrate how they are put to work in OntoVPA; we also outline OntoVPA's architecture. Next, we illustrate the core DM capabilities of OntoVPA in a *Point-Of-Interest Recommender VPA*. We analyze an example dialogue with OntoVPA in this domain and explain how ontologies, reasoning, and ontology-based rules enable OntoVPA's conversational, deep understanding. We then discuss OntoVPA's ontology in more detail; OntoVPA employs a generic, upper level ontology for dialogue management and domain models. Specific VPA domains are reusing, customizing and extending these upper level ontologies. We hence discuss the upper level ontology in some detail, as it represents core notions of the DM model. The runtime dynamics of the dialog (request and answer processing, turn taking, …) are implemented by ontology-based rules, and we discuss our DM-specific rule language. We conclude with a summary and outlook for future work.

2 Ontologies, Rules, and Architecture

In the following, we are using a *Restaurant Recommender VPA domain*. As one would expect, a *domain-specific restaurant ontology* likely contains *concepts* ($\approx$ classes, types, …) such as *Restaurant*, *ItalianRestaurant*, …; *relations* such as *servesFood*; *attributes* such as *hasAddress*, as well as specific *instances* of these classes, with *relationships* holding between them. Moreover, there are relationships between classes, i.e., *subclassOf*, *disjointWith*, *equivalentWith* etc.

In this domain, let us assume a restaurant of class/type *ItalianPizzaRestaurant* has been suggested to the user in response to a "*Find me an Italian restaurant!*" request. In subsequent dialogue steps, the user might refer back to this recommended restaurant, with follow-up questions such as "*What is the rating of the pizza place?*" or "*What is the price level of the Italian restaurant?*"—if the domain ontology states that the classes *ItalianRestaurant* and *PizzaPlace* are superclasses (hypernyms) of the *ItalianPizzaRestaurant* class, then these anaphora can be resolved by introspection of the dialogue history. OntoVPA contains generic co-reference/anaphora resolution rules. These ontology-based rules consider the class hierarchy of the ontology (the so-called *taxonomy*) and are aware of hypernyms, hyponyms, and synonyms (also compare [14]).

An expressive *ontology-based dialog representation* is a prerequisite for this kind of ontology-based anaphora resolution. The dialogue representation also needs to be "reflexive" in the sense that it is not sufficient to only remember what the user has said—the VPA/DMS also needs to remember its own utterances/dialogue steps, e.g., the suggested *ItalianPizzaRestaurant*.

The classes and relation types defined in the *static dialogue ontology* are instantiated in the *dynamic dialogue representation* at runtime. The *vocabulary* (= sets of classes, relations, and attributes) in the dialogue ontology is inspired by *speech act theory* [15]. Ontology-based rules over these representations then implement the *dynamics of the dialog by computing answers/responses.*

We are using the W3C standard OWL 2 [16] for the ontology language, and have implemented a custom ontology-based rule engine, based on the SPARQL 1.1 RDF query language [17]; the utilized Jena SPARQL engine [18] operates on the *deductive closure* induced by the OWL ontology, and is hence aware of the ontology consequences (unlike ordinary RDF query languages). Mature implementations of OWL 2 and SPARQL exists [18, 19], as well as modeling workbenches [20]. The use of W3C standards facilitates customer acceptance, and "off the shelf" ontologies are only readily available in standard formats such as OWL.

In a nutshell, ontologies facilitate the following in OntoVPA:

Domain model representation: The classes (types), relationship types, instances, and relations of the VPA domain. Often, we wish to reuse and extend well-known upper level ontologies, e.g., *Schema.org* [21]. Using standard OWL 2 Description Logic (DL) terminology [16, 22], the *classes* ($\approx$ *concepts, types, ...*) and *object and datatype properties* ($\approx$ *properties, slots, attributes, relations, parameters, ...*) constitute the *TBox* (*terminological box*) of the ontology, representing the *vocabulary* of the domain, whereas the actual instances of these classes and relationships are kept in the *ABox* (*assertional box*).

The dialogue ontology is a TBox—its classes and properties are inspired by *speech act theory* [15]. It contains classes for different kinds of dialogue steps, such as the *UserIntent* class, which is a subclass of the *UserRequest* class, which is a special *DialogUserStep*, and so on. It also contains classes for representing the system's responses. The dialogue ontology hence contains all classes, attributes, and relations (vocabulary) that are required for the actual runtime representation of the dialogue. This vocabulary is then instantiated at runtime in

The dialogue representation, which is an ABox—here, the classes of the dialogue ontology are instantiated. The instances represent the dialog steps (i.e.,

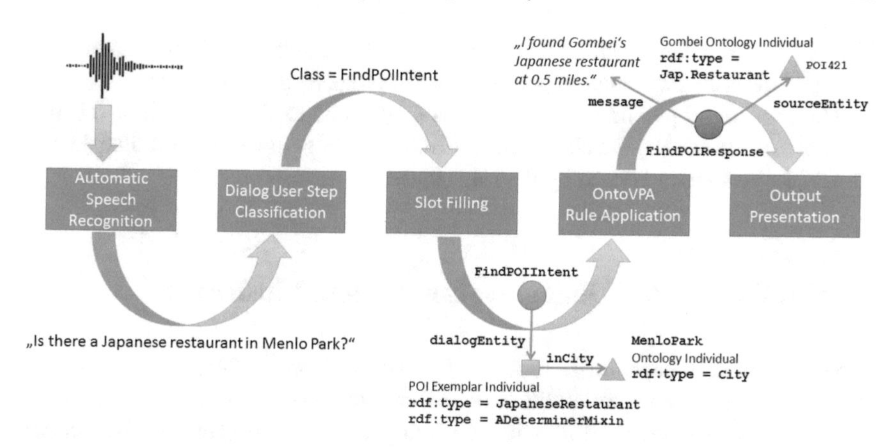

Fig. 1 OntoVPA's processing pipeline. *Automatic Speech Recognition (ASR)* turns the audio signal into text. The text is then classified as some kind of *DialogueUserStep*, e.g., a specific *UserIntent* such as *FindPOIIntent*. The ontology specifies parameters/slots for dialogue step classes, and the *slot filler* "fills in" these slots. The instantiated and slot-filled intent class then refers to individuals, classes, and relations from the ontology. Collectively, the *classifier and slot filler* act as a *semantic parser.* The constructed dialog step instance is inserted into the dialogue ABox, and the *rule engine* is invoked to compute OntoVPA's response, which is then presented (using different modalities) by the Output Presentation module

steps of the user and steps of the VPA system, requests and responses, …), and relationships between them represent causal and temporal dependencies between steps. There are also relationships to individuals from domain-specific ABoxes, so-called data source ABoxes.

Reasoning is applicable in a number of areas and offers the following potentials:

Deep domain knowledge and expertise: A VPA with deep expertise in a domain (a VPS) requires extensive and profound domain knowledge, and workflows and reasoning procedures are becoming more complex. Automatic reasoning by means of highly expressive ontology-based rules can solve complex domain-specific problems in a declarative, reusable, and human comprehensible (i.e., explainable) way.

Response computation: Ontology-based rules can be used to compute the actual utterances of the VPA, combining the best of conceptual, human-friendly knowledge representation with the expressive and human comprehensible (i.e., explainable) inferential power of rules.

Handling polysemy and ambiguity of natural language in dialogues: Ontologies provide a means to deal with the polysemy of natural language (NL). An ontology can structure the domain-specific vocabulary ("lexicon") in terms of semantic relationships between domain concepts/terms, such as hypernym/hyponym, synonym/antonyms, and so on. This knowledge can then be exploited for a variety of *context-specific NL interpretation and NL understanding tasks* [14], e.g., anaphora resolution, *semantic role labeling,* etc.

We will illustrate these potentials by means of an example domain in greater detail now, and describe how these capabilities are implemented in OntoVPA.

The processing pipeline of OntoVPA is shown and explained in Fig. 1 (cf. [4]). The ontology is used by the classifier, slot-filler, and rule engine. The ontology-based SPARQL rules are only used by the rule engine.

3 Ontology-Based Dialogue Management Illustrated

What are the core capabilities required for conversational VPAs/VPSs (cf. [4, 14, 23])? In our experience, in addition to anaphora/coreference resolution, the list of requirements includes the following: realizing when a user request (also: *intent*) is fully specified and ready for execution (i.e., all *required parameters/slot fillers* are specified); inquiring about missing required parameters/slot fillers; interpreting *arbitrary user input* in the context of the current dialogue (what does "In Palo Alto!" mean in the context of the current dialogue?); being able to maintain sub-dialogues/sub-workflows (e.g., the *book a business trip* dialogue requires a *book a hotel room* sub-dialogue); being able to determine and continue with an open, not yet fully executed (sub-) dialogue from the agenda of open (sub-)dialogues; to cancel an open (sub-) dialogue; support for refining, generalizing, deleting or overwriting slot values of previously executed requests (intents), and for reexecuting such a modified dialogue step; and, last but not least, the ability to recognize and disambiguate ambiguous input (does "Stanford" refer to the city, or the college?) in the context of the current dialogue.

OntoVPA supports all of these, and more. We will now illustrate some of these core capabilities with the example dialogue from Fig. 2 in the *Point-of-Interest (POI) Recommender* domain, which is a generalization of the Restaurant Recommender domain (i.e., it offers a richer domain ontology).

After an initial greeting from the system in **Step 1**, the user initiates the dialogue in **Step 2** by asking for a *Chinese Restaurant in Menlo Park.* The user's request is recognized/classified as a *FindPOIIntent*, a subclass of *UserIntent*. *DialogUserStep* classes are defined in the *dialogue ontology*. The dialogue ontology defines the vocabulary instantiated in the dialogue ABox, as illustrated in Fig. 3. The user's utterances (often, instances of *UserRequest* or *UserIntent*) are instantiated and inserted into the dialogue ABox by the *semantic parser*, whereas OntoVPA's utterances (usually *SystemResponses*) are created by OntoVPA's ontology-based rule engine. OntoVPA keeps track of the current dialogue user step and current dialogue system step by annotating the corresponding instances in the dialogue ABox with so-called *control markers*, i.e. *CurrentDialogUserStep* and *CurrentDialogSystemStep*.

The *domain ontology* contains a taxonomy of POI classes, such as *Chinese-Restaurant*; a subclass (= kind) of *AsianRestaurant*, which is a kind of *Restaurant*, which is a kind of *POI*, and so on. In addition, there are classes such as *City*. The actual POIs (instances of POI classes) are stored in a *domain data source*, which is an OWL (RDF) ABox—the *POI database*. These POIs have their typical attributes

No.	User Step (*DialogUserStep*)	OntoVPA Step (*DialogSystemStep*)
1	(none)	Hello, how can I help you?
2	Is there a Chinese Restaurant in Menlo Park?	Sorry, I can't find any Chinese Restaurant in Menlo Park. Do you want me to find something similar?
3	Okay!	I found Gombei Japanese restaurant in Menlo Park, at 0.34 miles.
4	Can you show it on the map?	I am assuming with 'it' you are referring to Gombei Japanese restaurant. Now showing it on the map.
5	Is there a library close to the Asian restaurant?	Yes, the Menlo Park library is close to Gombei Japanese restaurant, at 0.3 miles.
6	How about a Pizzeria in Menlo Park?	I found Applewood Pizzeria in Menlo Park, at 0.2 miles.
7	In Palo Alto!	I found Olive Garden in Palo Alto, at 1.4 miles.
8	Is there parking close to the Pizza Place?	Do you mean Olive Garden?
9	No	Applewood Pizzeria has a parking lot.

Fig. 2 Example Dialogue with the *Point-of-Interest (POI) Recommender* in OntoVPA

(properties), i.e., name, address, geographic coordinates, and so on. The POI Abox also contains cities such as *MenloPark*. For the sake of simplicity, POI instances refer to the cities in which they are located via the *inCity* object property (slot).

Along with the *FindPOIIntent*, the semantic parser also *asserts* (inserts, adds) a "fresh" *ChineseRestaurant* instance as a filler of the *dialogueEntity* slot (object property) into the dialogue ABox. This *ChineseRestaurant* POI instance can be thought of as a "query-by-example POI" for the *FindPOIIntent* query (see Sect. 5). The parser also realizes that "Menlo Park" refers to the *MenloPark City* individual in the POI database, and fills it in as slot value of the *inCity* slot on the fresh *ChineseRestaurant* instance. Slot-filling utilizes the ontology in order to identify the right slot (or likely slot candidates in case of ambiguity)—here, the *inCity* property is selected as the slot for the *MenloPark City* instance, because it is an instance of the slot's *range class* (*range*(*inCity*) = *City*).

OntoVPA now realizes that the *FindPOIIntent* is *completely specified*—all required parameters are given. Hence, it is ready for execution, and the semantic search over the POI database is performed. In this example, OntoVPA does not find a matching *ChineseRestaurant* in *MenloPark*, and it takes the initiative by proactively asking the user if the query should be *generalized.* Now, a *YesOrNoAnswer* is expected from the user. For both possible answers, OntoVPA has set up corresponding positive and negative *continuation requests;* the negative continuation is a *GreetingIntent*, whereas the positive continuation is a *FindPOIGeneralizedIntent*. The parameters required for the latter intent are copied over from the previous *FindPOIIntent*. Hence, depending on whether a *YesUserResponse* or a *NoUserResponse* is received, the corresponding prepared continuation intent is triggered automatically by OntoVPA. In **Step 3**, the *FindPOIGeneralizedIntent* is triggered based on the user's "Okay!" response, which is classified as a *YesUserResponse*.

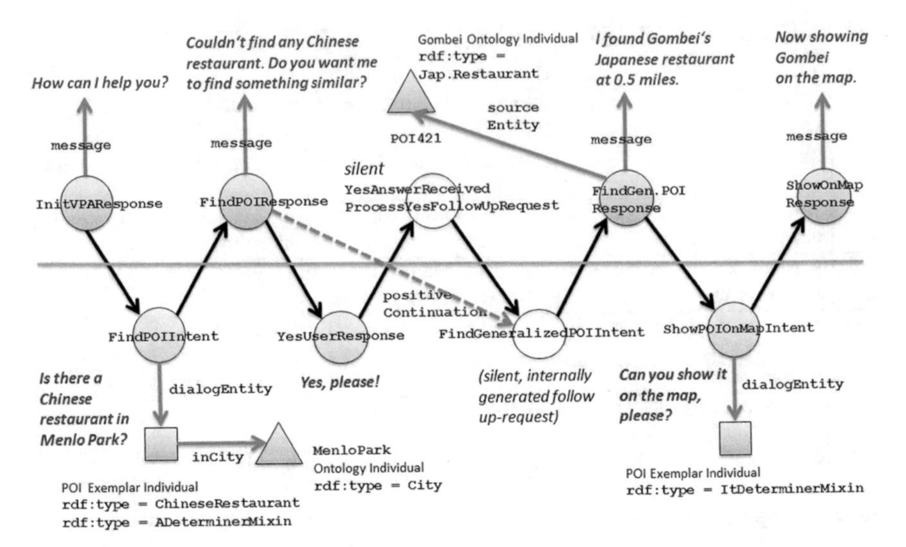

Fig. 3 Illustration of the Dialogue ABox after Step 4 from Fig. 2. Grey circles visualize OntoVPA's *SystemDialogueSteps*. Blue circles visualize the user's *UserDialogueSteps*. Blue shapes are created by the semantic parser; blue rectangles are instances of domain classes (e.g., *ChineseRestaurant*). Yellow triangles visualize ontology individuals from some data source (*MenloPark*). White circles are created programmatically as follow up requests by the rule engine

The *FindPOIGeneralizedIntent* employs a *relaxed semantic matching condition* which uses the taxonomy to compute a *semantic similarity measure* between the given query-by-example POI, and the actual source candidate POI. For example, a *JapaneseRestaurant* is more similar to a *ChineseRestaurant* than a *SteakHouse*, given that the least common subsumer (least common superclass) between *ChineseRestaurant* and *JapaneseRestaurant* is *AsianRestaurant*, but only *Restaurant* between *ChineseRestaurant* and *Steakhouse* (hence, they are less similar). Using generalized semantic search, a matching *JapaneseRestaurant* POI is found and linked to from the dialogue ABox by using the *sourceEntity* slot of the generated *FindPOIGeneralizedResponse* instance.

In **Step 4**, "Can you show *it* on the map?", OntoVPA realizes that *it* refers to the most recently discussed POI—like the already discussed *FindPOIIntent*, the *MapPOIIntent* has a *dialogueEntity* slot, with a local range restriction class *POI* (individuals can instantiate multiple classes in OWL). The *ItDeterminerMixin* anaphora resolution rule now identifies the most recently discussed entity from the dialogue that satisfies the types of the *MapPOIIntent*'s *dialogueEntity* slot filler, i.e., the most recent *POI* that occurred in the dialogue history. By traversing the dialogue history backward, it finds the *sourceEntity* filler of the previous *FindGeneralizedPOISystemResponse*. Hence, the retrieved Japanese restaurant is identified as the referent of the *it* anaphora. OntoVPA contains anaphora resolution rules for *it*, *the*, *a*, *his*, *her*, and so on. With the exception of *it*, they also instantiate and consider the given *class*, e.g., like in *the Asian Restaurant*.

Anaphora resolution involving the *TheDeterminerMixin* is illustrated in **Step 5**; there, it is realized that a POI instance of the class *JapaneseRestaurant* is also an instance of the *AsianRestaurant* superclass.

To demonstrate disambiguation, we are adding some more dialogue objects to the discourse, by requesting a *Pizzeria* in *MenloPark* in **Step 6**. In **Step 7** it is demonstrated how arbitrary input can be interpreted in the current context of the dialogue—based on the dialogue history, OntoVPA understands that the most likely user intent behind the ambiguous "In Palo Alto!" utterance is to *modify and reexecute the previous intent,* i.e., to look for a *Pizzeria* in *PaloAlto* instead of *MenloPark*. This introduces yet another *Pizzeria* instance, this time in Palo Alto.

Given that it is not self-evident what "In Palo Alto!" means without the context of the full dialogue, the semantic parser cannot instantiate a very specific *DialogueUserStep* or *UserIntent* class here. Instead, a generic high-level *ArbitraryUserInput* dialogue step with a *PaloAlto City* individual filler of the *inCity* slot is created, given that the proposition "in" maps to the *inCity* slot. A context-specific dialogue rule then processes the *ArbitraryUserInput* dialogue step—by looking at the *previousDialogueUserStep FindPOIIntent*, OntoVPA now suspects that the *inCity* slot value of the previous *FindPOIIntent* shall be overwritten with the given one (i.e., *MenloPark* be replaced with *PaloAlto*), and the so-modified *FindPOIIntent* be *reexecuted.*

In **Step 8**, there are now two pizzerias in the dialogue ABox—as the *PizzaPlace* class is a synonym for the *Pizzeria* class, "the Pizza Place" is now ambiguous and could refer either to the Pizzeria in Palo Alto, or to the one in Menlo Park. One disambiguation strategy (out of several available) is to ask for clarification, as illustrated in Fig. 2. OntoVPA also uses inference in **Step 9** to realize when the anaphora has been disambiguated, and when the original *FindPOIIntent* from Step 8 can be executed (notice that the exemplar *POI* instances also have an optional *nearBy* attribute, to which "close to" maps).

None of the just discussed DM strategies are hard coded—strategies can be changed flexibly by adding, enabling, or disabling generic DM rules in OntoVPA's *upper-level generic rule layer,* or by overwriting and/or specializing generic rules with domain-specific rules in the *domain-specific rule layer.*

4 OntoVPA's Upper Ontology and Modeling in OWL

OntoVPA's upper level ontology plays a key role for organizing and categorizing the different vocabularies into domain- and dialogue-specific parts, depending on role, and for categorizing *DialogueUnits* into certain speech acts. The upper level ontology contains two main branches: the *upper level dialogue ontology* branch, and the *upper level domain ontology* branch. Important root classes are:

DialogueStep: Root class of the dialogue ontology. Children of *DialogueStep* are: *DialogueUserStep*, *DialogueSystemStep*, *Request*, *Response*. Further down:

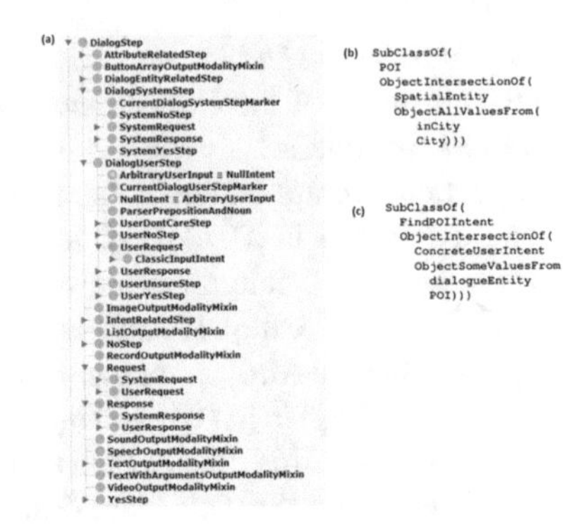

Fig. 4 Subclasses of *DialogStep* **(a)**, The *POI* Domain Class and *FindPOIIntent* Dialogue Class in OWL 2 **(b)**, **(c)**

DialogueUserRequest, *DialogueUserResponse*, *DialogueSystemRequest*, and *DialogueSystemResponse*. An important *DialogueUserRequest* subclass is *UserIntent*; all domain-specific intents such as *FindPOIIntent* specialize it. Arbitrary (highly dialogue context-dependent input) can be represented with *ArbitraryUserInput*. Most intents have a corresponding *SystemResponse* class, e.g., *FindPOIResponse*.

DomainNotion: Root class of the domain ontology. Only a few high-level notions such as *Entity*, *Event*, *TemporalThing*, and *SpatialThing*, are present. We also include Schema.org [21]. For example, the *AsianRestaurant* class will specialize *Schema.org/Restaurant*, and *ChineseRestaurant* will specialize *AsianRestaurant*.

DialogueControlMarkers: are used to control DM of the rule engine. We already mentioned *CurrentDialogueUserStep* and *CurrentDialogSystemStep*.

In addition to the *class hierarchy* (*see* Fig. 4a), the upper ontology also contains an *object property hierarchy* and a *datatype property hierarchy*. These property hierarchies mirror the class hierarchy closely (all hierarchies support multi-inheritance): there are root properties corresponding to the three main root classes. Important *control* properties are *nextStep*, *previousStep*, and *finalSystemResponse*; these correspond to edges in the dialogue ABox graph of Fig. 3. Control properties are often used in combination with *DialogueControlMarker*s control vocabulary to control and instruct OntoVPA's rule engine, for dialogue branching, looping, follow-up continuations based on the user's response, etc.

Modeling properties of domain classes and intents in OWL [16, 22] *seems* to be straightforward—distinguishing between *required* and *optional* parameters/properties turns out to be challenging though, mainly due to the *Open World Assumption (OWA)* in OWL. We argue that an *epistemic semantics* is desirable for specifying required properties of *DialogUserStep*s or *UserIntent*s. For example, the *FindPOIIntent* cannot be executed unless the *required dialogueEntity* is *known* (hence, *epistemic* semantics). Due to the OWA, simply declaring a property on a class by using an existential restriction of the form $\forall x : FindPOIIntent(x) \Rightarrow$

$\exists y : dialogueEntity(x, y)$ is insufficient: Given a dialogue ABox such as $\{FindPOIIntent(userIntent1)\}$, where the *dialogueEntity* filler is not given explicitly, the OWL reasoner will infer that *some dialogueEntity* POI instance exists. However, we would like to require that this instance should be explicitly given by the user. To complicate things even further, the *POI* domain class has an *optional inCity* property. From a logical point of view, under the standard OWA, every property is optional on an OWL class or instance, as long as it doesn't cause an inconsistency.

We are distinguishing required from optional properties as follows. A *required* property (on a *DialogueSteps*) is declared with an *existentially quantified* axiom (using `someValuesFrom` in OWL), as just discussed: $FindPOIIntent(x) \Rightarrow \ldots (\exists y : dialogueEntity(x, y)) \ldots$. However, OntoVPA'S DM system will, on *DialogueSteps*, always interpret such properties under the stricter epistemic "must be known" semantics, and inquire in case of missing slot values for these. *Optional* properties are declared by means of universal quantifiers such as $\forall x : POI(x) \Rightarrow \ldots \wedge (\forall y : inCity(x, y) \Rightarrow City(y)) \ldots$ (`allValuesFrom` in OWL). This is also illustrated in Fig. 4b, c; *inCity* is optional on *POI*, and *dialogueEntity* is required on *FindPOIIntent*.

5 Rules for Dialogue Management and Workflows

Ontology-based rules are used to compute OntoVPA's responses, to implement DM strategies, domain workflows, and the majority of the domain-specific "application logic". Based on the SPARQL 1.1 RDFs query language [18] implemented by Jena [18], we have implemented a DM-specific rule language. The employed Jena SPARQL engine is aware of the *inferred triples* implied by the ontology axioms, and hence offers expressive OWL reasoning capabilities. SPARQL `construct` queries are used to dynamically augment the dialogue ABox.

The rule engine offers versatile and flexible rule application and conflict resolution strategies, if more than one rule is applicable in a given dialogue situation. For example, conflict resolution can be performed based on the specificity of a rule, based on a numeric preference, or based on an explicitly specified defeasibility relation. The briefly mentioned *DialogueControlMarkers* play a crucial role in controlling and "advising" the rule engine. It is aware of the special semantics of the DM-specific control vocabulary and implements a discourse-specific semantics for them.

A simple rule that responds to a *GreetingIntent* looks as follows in OntoVPA. It plays back the fixed *GreetingSystemResponse* defined in the ontology:

```
@
Hello-Toplevel
Reply to a greeting intent. Answer strings are defined in ontology.
1
CONSTRUCT { ?o vpa:assertedType vpa:CurrentDialogueSystemStepMarker }
WHERE
 { ?i vpa:assertedType vpa:CurrentUserIntentMarker .
   ?i vpa:assertedType vpa:GreetingIntent .
   ?i vpa:finalExpectedSystemResponse ?o }
```

A set of such rules separated by "@" is specified in a `.sparql` (rule) file. Each rule has a name; rules with a `-Toplevel` suffix act as "daemons"—they are automatically (re)checked for applicability each time the dialogue ABox gets updated, and fired automatically by the DMS (after conflict-resolution). Non-daemon rules can be triggered by other rules, to act as *follow-up or continuation rules*. The third line (1) provides control information to the rule interpreter, such as rule priority, and precedence over other rules (defeasibility reasoning).

In the example rule just discussed, a SPARQL variable `?o` is bound to a *GreetingIntentSystemResponse*, which was inserted into the dialogue ABox by the rule engine together with the *GreetingIntent*. This is possible because the *GreetingIntentSystemResponse* class is declared as the *expected system response class* of the *GreetingIntent* class. The rule then simply annotates the pre-computed response individual with the *CurrentDialogueSystemStepMarker*, and the predefined answer string is retrieved and played back from the ontology.

We like to mention *three features of SPARQL 1.1 that are essential* for OntoVPA. The *first essential feature* is the ability to `construct` new nodes and assertions in the dialogue ABox by means of rules, using so-called "_:blank" nodes and SPARQL's `construct`. *Fixed utterances* such as the discussed greeting response are often insufficient—*templates* are needed. Within a template, it must be possible to refer to variable bindings, as well as to static `rdfs:label` strings from the ontology (which can be defined in different languages). SPARQL's `xfn:concat` string concatenation function is used to construct the new answer string. Hence, the *second essential feature* of SPARQL for OntoVPA are *(user-defined) extension functions,* such as `xfn:concat`. Similar to *procedural attachments*, they can invoke arbitrary Java code during rule execution.

Continuing with the *GreetingIntent* example rule, instead of playing back the pre-constructed answer string from the *GreetingResponse* class, we can dynamically create a fresh *FinalResponse* instance with a blank node `_:o`, and construct a message from a template that includes the name of the current user:

```
CONSTRUCT {
  _:o vpa:assertedType vpa:CurrentDialogueSystemStepMarker .
  _:o vpa:assertedType vpa:FinalResponse .
  _:o vpa:assertedType vpa:TextOutputModalityMixin .
  _:o vpa:message xfn:concat("Hi ", ?name, "! How are you?") }
```

The *third essential feature* of SPARQL in OntoVPA is the ability to perform (a limited form of) *existential and universal second order quantification.* Consider the *FindPOIIntent* rule. This rule has to ensure that a candidate *SourcePOI* from the POI ABox fulfills all the requirements expressed by the exemplar *queryPOI*, i.e., it must have (at least) all the properties of the exemplar POI (it *could* also have more specific properties and/or fillers, but we ignore this for the sake of brevity). This can be expressed as a second-order quantification over properties P: $\forall P : \forall val : P(queryPOI, val) \Rightarrow P(sourcePOI, val)$. Notice that *sourcePOI* can have more properties than required by the *queryPOI*. With some further refinements (we can restrict the quantification to *entityAttribute* subproperties), we can express this in SPARQL as follows (`?qentity` is the exemplar *query* POI, and `?sentity` is the *source* POI):

```
?qentity vpa:assertedType ?qtype .
?sentity rdf:type ?qtype .
?sentity rdf:type vpa:SourceEntity .
FILTER NOT EXISTS {
   ?qentity ?par ?parVal .
   ?par rdfs:subPropertyOf vpa:entityAttribute .
   FILTER NOT EXISTS {
     ?sentity ?par ?parVal } }
```

This implements a *generic semantic search*, for all kinds of `vpa:Source Entities`. Many of OntoVPA's generic DM capabilities are implemented succinctly and declaratively by similar *higher-order rules*.

6 Conclusion and Outlook

We have presented the ontology-based DMS OntoVPA, and illustrated its underlying techniques. During the last 3 years, OntoVPA has been successfully deployed to 4 different SRI customers, ranging from domains as different as Beauty Consultant, Shopping Assistant, Car Conversational System, to an Augmented Reality Tutoring & Mentoring System.

One of these VPAs uses Japanese language (internally, English is being used). OntoVPA supports *multi-modal* input and output—different input modalities are represented by means of dedicated *inputModality* slots on the asserted *DialogUserSteps*, and dedicated *outputModality* values can be constructed on *SystemResponses*—`construct` is able to create arbitrarily complex (nested, cyclical, …) output structures.

In these projects, we have observed a significant reduction in VPA development time and costs, compared to previous VPA projects @ SRI that did not use OntoVPA. The exact numbers and evaluation are subject to future research.

We have based OntoVPA and its dialogue representation on formal, explicit, standardized, and expressive symbolic knowledge representations. This has the benefit of being transparent and reusable—the dialogue history can be introspected, reasoned about, visualized (for example, using OWL visualizers); shared, persisted, resumed, etc. Moreover, its behavior is explainable.

The features and strategies for DM that are pre-defined by OntoVPA are powerful, yet flexible and relatively easy to change and extend if necessary. The system is highly expressive and is also capable to "emulate" a variety of simpler DMS architectures/paradigms—for example, it is straightforward to implement a finite state machine-based DMS. Moreover, OntoVPA is *reflexive* and *meta circular*—instead of encoding the DM strategies as high-order rules as just discussed, it is also possible to encode these strategies on an *instance level* in the ABox and create higher-order *meta interpreter DM rules* that then introspect and interpret the instance-level encoded DM strategies and execute them. We hence suspect that OntoVPA subsumes most of the existing DMS architectures on the market. We are musing about participating in a future *Dialogue State Tracking Challenge* to evaluate OntoVPA's performance [2]. We would also like to mention that OntoVPA can be licensed from SRI International.

References

1. Lison P (2013) Structured probabilistic modelling for dialogue management. University of Oslo, Diss
2. Williams J, Raux A, Ramach D, Black A (2013) The dialog state tracking challenge. In: Proceedings of the 14th annual meeting of the special interest group on discourse and dialogue (SIGDIAL)
3. Mark W, The rise of virtual specialists. https://www.sri.com/blog/deep-knowledge-and-rise-virtual-specialists. Accessed 2 Aug 2017
4. Lee C, Jung S, Kim K, Lee D, Lee GG (2010) Recent approaches to dialog management for spoken dialog systems. J Comput Sci Eng (JCSE) 4(1):122 (April 2010)
5. Gunning D, Chaudhri V, Clark P, Barker K, Chaw S, Greaves M, Grosof B, Leung A, McDonald D, Mishra S, Pacheco J, Porter B, Spaulding A, Tecuci D, Tien J (2010) Project halo update–progress toward digital aristotle. AI Magazine, AAAI Press (Oct 2010)
6. Wantroba E, Romero R (2014) A method for designing dialogue systems by using ontologies. In: Standardized knowledge representation and ontologies for robotics and automation. Chicago, USA, 18th Sept 2014
7. Pardal J, (2007) Dynamic use of ontologies in dialogue systems. In: Proceedings of the NAACL-HLT 2007 doctoral consortium. Association for Computational Linguistics, April 2007
8. Pardal J (2011) Starting to cook a coaching dialogue system in the olympus framework. In: Proceedings of the paralinguistic information and its integration in spoken dialogue systems workshop. Springer, pp 255–267
9. Chaudhri V, Cheyer A, Guili R, Jarrold B, Myers K, Niekarsz J (2006) A case study in engineering a knowledge base for an intelligent personal assistant. In: Proceedings of the 5th international conference on semantic desktop and social semantic collaboration
10. Liu G (2012) A task ontology model for domain independent dialogue management. Electronic Theses and Dissertations, University of Windsor, Paper 5412
11. Heinroth T, Denich D, Schmitt A, Minker W (2010) Efficient spoken dialogue domain representation and interpretation. In: Proceedings of the seventh international conference on language resources and evaluation (LREC 2010)
12. Ultes S, Dikme H, Minker W (2016) Dialogue management for user-centered adaptive dialogue. In Situated dialog in speech-based human-computer interaction, 2016. Springer International Publishing, pp 51–61
13. Sonntag D, Huber M, Möller M, Ndiaye A, Zillner S, Cavallaro A (2010) Design and implementation of a semantic dialogue system for radiologists. In: Haffner KA (eds) Semantic web: standards, tools and ontologies. Nova Science Publishers
14. Milward D, Beveridge M, Ontology-based dialogue systems. In: Proceedings of the 3rd workshop on knowledge and reasoning in practical dialogue systems (IJCAI 2003) (Aug 2003)
15. Searle JR (1969) Speech acts. An essay in the philosophy of language. Cambridge University Press, 2 Jan 1969
16. OWL 2 web ontology language-structural specification and functional-style syntax, 2nd edn. https://www.w3.org/TR/2012/REC-owl2-syntax-20121211/. Accessed 2 Aug 2017
17. SPARQL 1.1 Query Language, W3C. https://www.w3.org/TR/sparql11-query/. Accessed 2 Aug 2017
18. Apache Jena. https://jena.apache.org/. Accessed 2 Aug 2017

19. Haarslev V, Hidde K, Möller R, Wessel M (2012) The RacerPro knowledge representation and reasoning system. Semant Web J 3(3):267–277
20. Protégé 5 Visual Ontology Modeling Environment. http://protege.stanford.edu/. Accessed 2 Aug 2017
21. Schema.org. http://schema.org/. RDFs exports under. http://schema.rdfs.org/
22. Baader F, Calvanese D, McGuinness D, Nardi D, Patel-Schneider PF (2003) The description logic handbook: theory, implementation, and applications. Cambridge University Press
23. Vertan C, Hahn WV (2017) Project "Spoken Dialogue Systems", Seminar slides. https://nats-www.informatik.uni-hamburg.de/pub/DIALSYS/VeranstaltungsMaterial/DialogueManagement.pdf. Accessed 13 April 2017

Regularized Neural User Model for Goal-Oriented Spoken Dialogue Systems

Manex Serras, María Inés Torres and Arantza del Pozo

Abstract User simulation is widely used to generate artificial dialogues in order to train statistical spoken dialogue systems and perform evaluations. This paper presents a neural network approach for user modeling that exploits an encoder-decoder bidirectional architecture with a regularization layer for each dialogue act. In order to minimize the impact of data sparsity, the dialogue act space is compressed according to the user goal. Experiments on the Dialogue State Tracking Challenge 2 (DSTC2) dataset provide significant results at dialogue act and slot level predictions, outperforming previous neural user modeling approaches in terms of F1 score.

Keywords User simulation · Dialogue systems · Deep learning · Regularization

1 Introduction

Developing statistical Spoken Dialogue Systems (SDS) requires a high amount of dialogue samples from which Dialogue Managers (DM) learn optimal strategies. As manual dialogue compilation is highly resource demanding, an usual approach is to develop an artificial user or User Model (UM) from a small dataset capable of generating synthetic dialogue samples for training and evaluation purposes [19]. UMs are designed in a way that they receive an input from the DM and return a coherent response. A consistent model is expected to maintain coherence throughout the dialogue and to imitate the behavior of real users. Also, some degree of variability is desired in order to generate unseen interactions.

M. Serras · A. del Pozo
Speech and Natural Language Technologies, Vicomtech Research centre, Donostia-San Sebastian, Spain
e-mail: mserras@vicomtech.org

A. del Pozo
e-mail: adelpozog@vicomtech.org

M. I. Torres (✉)
Speech Interactive Research Group, Universidad del País Vasco UPV/EHU, Leioa, Spain
e-mail: manes.torres@ehu.es

M. Eskenazi et al. (eds.), *Advanced Social Interaction with Agents*, Lecture Notes in Electrical Engineering 510, https://doi.org/10.1007/978-3-319-92108-2_24

There have been several user modeling proposals in the literature. Initial approaches [6, 13, 14] used N-grams to model user behavior, but were not capable of capturing dialogue history and, thus, lacked coherence. Subsequent efforts to induce more coherent UMs were proposed by [17, 20]. However, these methods often required a large amount of hand-crafting to infer the dialogue interaction rules.

Several statistical UM approaches have tried to reduce the amount of manual effort required while maintaining dialogue coherence. In [15] Bayesian Networks were used to explicitly incorporate the user goal into the UM. A network of Hidden Markov Models (HMM) was proposed in [5], each HMM representing a goal in the conversation. A hidden-agenda where the user goal is predefined as an agenda of constraints and pieces of information to request to the system and updated at each dialogue turn was presented in [18]. Other approaches have proposed the use of inverse reinforcement learning [2], exploiting the analogies between user simulation and imitation learning.

Recently, a sequence-to-sequence neural network architecture has been proposed [12] for user modeling. Taking into account the whole dialogue history and the goal of the user, this method predicts the next user action as a sequence decoding of dialogue acts. Despite proven to be a promising approach, it suffers from data sparsity when it comes to represent dialogue acts at slot value level.

This paper proposes to model the user as an ensemble of bidirectional encoder-decoder neural networks. The dialogue history is encoded as a sequence instead of a single vector to avoid the information loss caused by compression [3]. Before the decoding process, an additional layer is used to learn regularization parameters that are applied to the encoded sequence in order to improve the generalization of the model. Each user dialogue act is trained in an independent network and an ensemble is constructed by joining all expert networks to predict the user action for each turn. In order to address the data sparsity problem, both system and user dialogue act representations are compressed at slot value level according to the user goal. This representation allows the slot level information to be included in the network during the training process, and thus, represents the dialogue interaction logic with finer granularity.

The paper is structured as follows: Sect. 2 introduces goal-oriented SDS and explains how dialogue act representations are compressed according to the goals set in the dialogue scenario. Section 3 describes the proposed neural network architecture in detail. Section 4 presents the experiments carried out on the DSTC2 dataset. Finally, Sect. 5 summarizes the main conclusions and sets guidelines for future work.

2 Compressing Goal-Oriented Dialogue Acts

Statistical approaches to dialog management require a large amount of dialog samples to train the involved models. Human-to-human dialogues are generally used to train open domain dialogue systems. However, for goal-oriented human-machine interaction a common practice to obtain controlled dialogue samples is to assign the

Table 1 Example scenario of a restaurant domain corpus and its goal representation

Description	You are looking for a restaurant in the north part of town and it should serve international food. Make sure you get the address and phone number
Constraints	*Food: International*
	Area: North
Requests	*Address*
	Phone

users a scenario to fulfill through the dialogue [8, 10]. Such scenario may contain diverse goals to complete by the user and other relevant information for the upcoming interaction.

As explained in [18], the dialogue goal $G = (C, R)$ can be represented as a set of constraints C to inform and values to request R that the user needs to fulfill through the interaction. Table 1 shows a dialogue scenario given in the Dialogue State Tracking Challenge 2 (DSTC2) corpus used in the experimental section of this paper, where $C = (food = international, area = north)$ and $R = (address, phone)$ are given explicitly. The constraints and requests of the goal have a direct correlation with the user's intention at semantic level. User-system interactions are usually represented through dialogue acts (DA) [1, 4, 7], denoted by intention tags (e.g. *inform, request, confirm*) which can contain information objects known as slots, with their corresponding values.

The following example shows an utterance annotated using the dialogue act schema: *inform* and *request* are the dialogue acts; *food, area, address* and *phone* are slots while *international* and *north* are slot values. Note that there can be dialogue acts without slots and that not all slots need to have a specific value.

Utterance: I want a restaurant that serves international food in the north. Give me it's address and phone number

DA Representation: *inform (food=international, area=north) & request (address, phone)*

The main problem of the dialogue act representation is the huge amount of possible slot values. For example, in the DSTC2 corpus, the system can inform of more than 90 food values and 100 restaurant names. Assuming the slot values returned by the system are relevant to the user only if they match a constraint set in the dialogue scenario, every slot value can be replaced with an *is_goal* or *not_goal* token, depending on whether or not they match the given constraint values. Following this assumption, the possible values of the food slot can be reduced from more than 90 to only two.

Table 2 shows how the slot values of an interaction represented at dialogue act level are compressed according to the user goal constraints given in Table 1. As it can be seen, this assumption has a direct impact in the dialogue act representation schema,

Table 2 Dialogue act interaction example, compressed according to the user goal

Original interaction	Compressed interaction
System: request(area)	System: request(area)
User: inform(area = north, food = international)	User: inform(area = is_goal, food = is_goal)
System: expl-conf(food = italian)	System: expl-conf(food = not_goal)
User: negate( )&inform(food = international)	User: negate( )&inform(food = is_goal)
System: offer(restaurant)&inform(area = north, food=international)	System: offer(restaurant)&inform(area = is_goal, food=is_goal)
User: request(address, phone)	User: request(address, phone)

narrowing down each slot to just two values. As a result, slot value level information can be included in dialogue act representations for user modeling purposes, avoiding excessive sparsity and with small information loss.

3 Regularized Bi-directional LSTM User Model

The neural network architecture proposed for user modeling is a bidirectional encoder-decoder with a regularization layer. It encodes the dialogue history in a sequence both forward and backward and exploits a regularization mechanism to set the focus only on the relevant sections of the encoded sequence.

As shown in Fig. 1, the input to the network is a concatenation of the user goal set in the dialogue scenario G and the sequence of system dialogue acts until the current turn t, $DA_s^t = (da_s^0, da_s^1, \ldots, da_s^t)$. G is represented as a 1-hot encoding of the slots given as constraints and requests in the dialogue scenario. The output of the network is a prediction of the user dialogue act at the current turn da_u^t. Note that while system dialogue acts change turn by turn, the initial goal representation remains the same throughout the dialogue.

The encoding layer is composed of a bidirectional Long Short Term Memory (LSTM) [9], whose output is the dialogue history encoded as $\mathbf{h_f}$ forward and as $\mathbf{h_b}$ backward.

The applied regularization mechanism requires to learn the weight vectors α for each row of the encoding matrix $H = [\mathbf{h_f}, \mathbf{h_b}]$. Being H_i the i-th row of the encoding matrix, the vector α_i is calculated as $\alpha_i = \sigma(W_a H_i)$, where W_a are the parameters of the Regularization Layer and σ the sigmoid function. Once H and α are known, the encoded sequence is regularized by the element-wise product as follows: $Reg = \alpha \odot H$. This operation will override the non-relevant values of the encoded sequence.

Decoding is then applied to Reg through another bidirectional LSTM, which outputs forward and backward decoding vectors d_f^t and d_b^t at turn t. These vectors

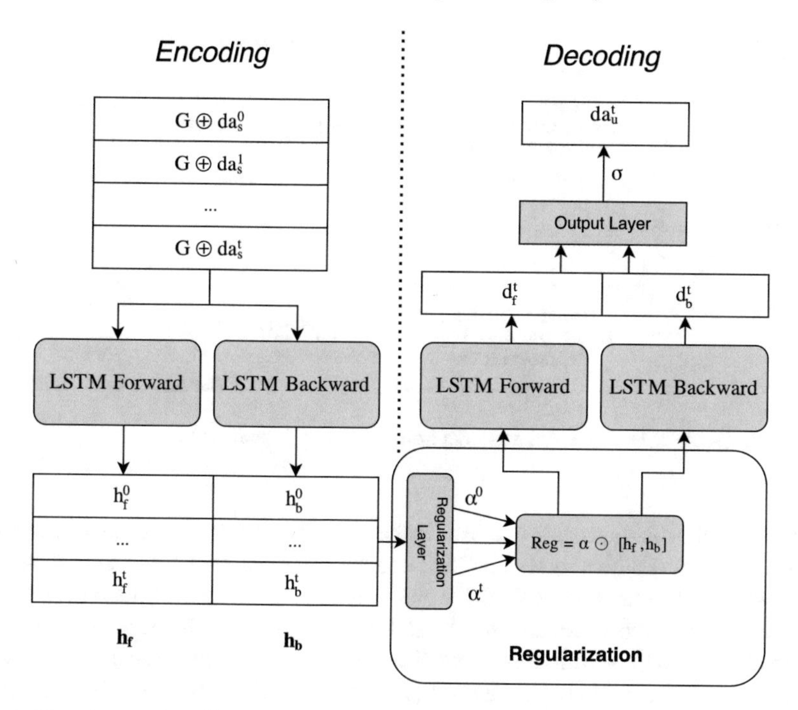

Fig. 1 Neural network architecture proposed for user modeling

are finally concatenated and processed by the output layer with a sigmoid activation function, from which the user dialogue act at the current turn da_u^t is predicted.

The proposed model uses an expert network for every possible user dialogue act, so the architecture in Fig. 1 is replicated for each dialogue act of the user. As a result, the final user model is an ensemble of networks, each of which predicts the slots of a specific user dialogue act as shown in Fig. 2. A dialogue act is triggered when the corresponding network of the ensemble returns any value above an individual threshold θ set in the development phase (e.g. θ_{inform} for the *inform* dialogue act). The final output is the combination of dialogue acts given by the ensemble of neural networks.

4 Experimental Framework

4.1 Dialogue State Tracking Challenge 2

The presented neural user model has been tested on the Dialogue State Tracking Challenge 2 (DSTC2) corpus. The second edition of the DSTC series [8] was focused

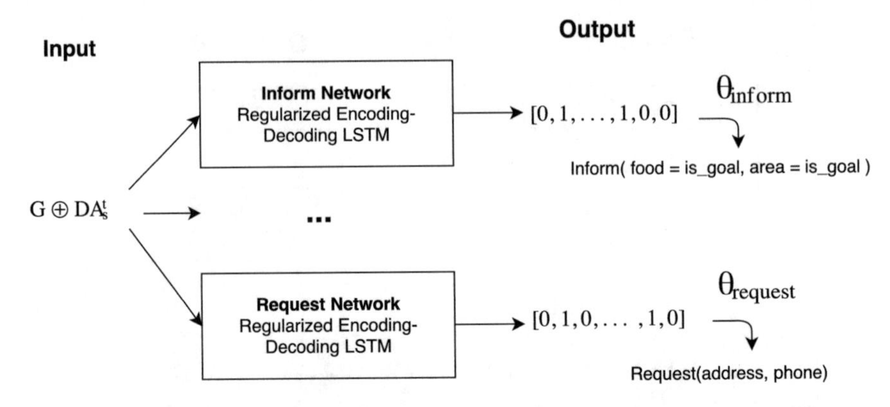

Fig. 2 Ensemble of Dialogue Act networks from which da_u^t is predicted

on tracking the dialogue state of a SDS in the Cambridge restaurant domain. For such purpose, a corpus with a total of 3235 dialogues was released.[1] Amazon Mechanical Turk was used to recruit users who would interact with a spoken dialogue system. Each user was given a scenario similar to that described in Table 1, which had to be completed interacting with the system. The goals defined in such scenarios followed the agenda approach of [18]. As a result, the constraints and requests of the user goal are explicitly annotated for each dialogue in the corpus. Table 3 summarizes the user dialogue acts of the DSTC2 corpus with their slots. Note that many dialogue acts do not have related slots and that the slots of the *Request* dialogue act have no value. Table 4 include all the informable slots in the DSTC2 corpus and some examples of their possible values.

The train/development/test set partitions of the corpus have been used to train, validate and test the proposed methodology. The development set has been used to set the thresholds ($\theta_{inform}, \ldots, \theta request$) and to control overfitting based on early stopping. The test set has been used to carry out final evaluation in terms of Precision, Recall and F1-score as in [5, 12, 16, 19]. These metrics allow comparing the dialogue acts of real and simulated users, measuring the behavior and consistency of the model.

4.2 Experiments and Results

This section describes how the ensemble of networks was trained on the DSTC2 corpus and shows the results achieved, both at dialogue act and slot levels.

[1]http://camdial.org/~mh521/dstc/.

Table 3 Dialogue acts that the user can trigger in the DSTC2 corpus

User dialogue act	Related slots
Acknolwedge	*Null*
Affirm	*Null*
Bye	*Null*
Confirm	*Area, Food, Price Range, Restaurant*
Deny	*Area, Food, Price Range, Restaurant*
Hello	*Null*
Help	*Null*
Inform	*Area, Food, Price Range, Restaurant*
Negate	*Null*
Repeat	*Null*
Request alternatives	*Null*
Request more	*Null*
Request	*Area, Food, Price Range, Restaurant, Phone, Address, Signature, Postcode*
Restart	*Null*
Silence	*Null*
Thankyou	*Null*

Table 4 Possible slot values for the *Inform, Confirm* and *Deny* dialogue acts

Informable slots	Possible values	Examples
Restaurant name	113	Nandos, Pizza Hut, …
Food type	91	Basque, Italian, European, …
Price range	3	Cheap, Moderate, Expensive
Area	5	North, west, south, east, centre

Training was done using mini-batch learning; having a dialogue N turns, the total batch is of size N and each input is the sequence of system dialogue acts until turn $t \leq N$. For gradient descent, the Adam [11] optimization method was used with a fixed step size of 0.001. No dropout nor weight penalties were employed. The loss function was computed using the squared error for multiple slot output dialogue acts (e.g. *inform, request*) and cross-entropy for single output dialogue acts (*bye, acknowledge)*. Each layer of every dialogue act network had 256 neurons. The individual threshold θ for each dialogue act network $(\theta_{inform}, \cdots, \theta_{bye})$ is set using the development set. In order to set the threshold's value for each dialogue act, a grid search is done to maximize the individual F1 score.

Table 5 shows the Precision, Recall and F1 score achieved by the proposed model with and without regularization on the DSTC2 development and test sets at dialogue act level. For comparative purposes, results presented by [12] in the first reported

Table 5 Overall results at dialogue act level

		Bi-directional LSTM	Regularized Bi-directional LSTM	Sequence-to-one [12]
DSTC2 Dev	Precision	0.69	0.70	–
	Recall	0.71	0.72	–
	F1	0.70	**0.71**	0.37
DSTC2 Test	Precision	0.68	0.71	–
	Recall	0.71	0.73	–
	F1	0.69	**0.72**	0.29

Table 6 Results for each user dialogue act

Speech act	Dev. set					Test set				
	Apparison	Predicted	Prec	Rec	F1	Apparison	Predicted	Prec	Rec	F1
Ack	0	0	0	0	0	9	0	0	0	0
Affirm	144	129	0.73	0.65	0.69	601	677	0.70	0.79	0.75
Bye	526	528	0.82	0.82	0.82	1169	1082	0.85	0.78	0.81
Confirm	39	0	0	0	0	46	0	0	0	0
Deny	4	0	0	0	0	4	0	0	0	0
Hello	18	0	0	0	0	39	0	0	0	0
Inform	1647	1696	0.80	0.82	0.81	4685	4456	0.85	0.82	0.83
Negate	68	62	0.51	0.47	0.49	261	217	0.46	0.38	0.42
Null	385	480	0.15	0.19	0.17	746	1335	0.10	0.18	0.13
Repeat	7	0	0	0	0	8	0	0	0	0
Reqalts	275	326	0.37	0.44	0.40	649	842	0.36	0.47	0.41
Reqmore	1	0	0	0	0	1	0	0	0	0
Request	1043	1093	0.77	0.81	0.78	2243	2299	0.80	0.82	0.81
Restart	3	0	0	0	0	7	0	0	0	0
Thankyou	510	531	0.79	0.82	0.81	1125	1101	0.80	0.78	0.79

neural user modeling approach are included, which outperformed previous bigram and agenda-based approaches. As it can be seen, the proposed method significantly improves the F1 score of the simulated user model. The improvement is justified by the increase of overall network complexity and the exploitation of compressed slot value level information for user dialogue act prediction. Also, the regularization mechanism slightly improves the generalization capability of the user model, so its regularized version has been used in the rest of the experiments.

Table 6 summarizes the results achieved for each user dialogue act. As it can be seen from the table, the proposed simulated user is capable of modeling high frequency dialogue acts with ease, but struggles when it comes to low frequency ones. Despite a neural network is trained using the whole corpus for each dialogue act, there are some cases where there is still not enough data to make any prediction.

Table 7 Overall results at slot value level

Dataset	Precision	Recall	F1
DSTC2 Dev	0.60	0.63	0.62
DSTC2 Test	0.60	0.64	0.62

Table 8 *Inform* dialogue act results at compressed slot value level

Inform acts	Slot value	Development set					Test set				
		Count	Predicted	Prec	Rec	F1	Count	Predicted	Prec	Rec	F1
Food	*is_goal*	509	719	0.54	0.76	0.63	1472	1970	0.60	0.80	0.69
	not_goal	299	204	0.39	0.26	0.31	809	775	0.45	0.43	0.44
Area	*is_goal*	423	454	0.67	0.72	0.70	1071	1042	0.70	0.68	0.69
	not_goal	21	0	0	0	0	115	0	0	0	0
Price range	*is_goal*	375	342	0.75	0.69	0.72	908	833	0.72	0.67	0.70
	not_goal	16	0	0	0	0	116	0	0	0	0
This	*Don't care*	231	308	0.61	0.81	0.70	767	898	0.59	0.69	0.64

Table 9 *Request* dialogue act results at slot level

Requested slot	Development set					Test set				
	Count	Predicted	Prec	Rec	F1	Count	Predicted	Prec	Rec	F1
Food	92	56	0.66	0.40	0.5	134	140	0.51	0.53	0.52
Area	40	21	0.71	0.38	0.49	113	36	0.75	0.24	0.36
Pricerange	90	62	0.64	0.44	0.52	115	93	0.62	0.50	0.56
Address	421	549	0.62	0.81	0.70	939	1154	0.68	0.84	0.75
Phone	426	498	0.64	0.75	0.69	986	999	0.72	0.73	0.72
Postcode	94	80	0.75	0.64	0.68	219	163	0.83	0.61	0.70
Signature	2	0	0	0	0	10	0	0	0	0

Table 7 shows overall results achieved at slot value level. As expected, performance decreases given the finer granularity of the task but still remains high considering the extra complexity involved.

Finally, Tables 8 and 9 show a more exhaustive evaluation of the two dialogue acts with highest impact on the DSTC2 corpus: *Inform* and *Request*, at compressed slot value and slot level respectively.

As it can be seen in the table, the simulated user achieves good results informing slot values set as goals in the initial scenario, but suffers from heavy degradation when it comes to inform those that are not defined as such.

In relation to the behavior of the user model with regard to the *Request* dialogue act, the correlation between high F1 scores and the requested slot occurrence is clear; the higher the slot occurrence, the better the F1 score.

5 Conclusions and Future Work

This paper has presented a neural user model for goal-oriented spoken dialogue systems. The proposed approach employs a sensible way to exploit slot level information without adding unnecessary sparsity to the representation. The ensemble of bidirectional encoding-decoding networks is capable of exploiting the full dialogue history efficiently and the regularization technique slightly improves the generalization capability of the model. These changes provide significant results both at dialogue act and slot level predictions, outperforming previous neural user modeling approaches in terms of F1 score.

Future work will require refining the presented architecture, so that it can model low-occurrence dialogue acts and slot values more precisely. The approach should also be tested on additional goal-oriented dialogue datasets. The proposed simulated user model will be compared against other user modeling approaches, when it comes to training and evaluating statistical spoken dialogue systems. In addition, having real users evaluate the generated policies shall provide useful insights about the modeling capabilities of the network.

Acknowledgements This work has been partially funded by the Spanish Minister of Science under grants TIN2014-54288-C4-4-R and TIN2017-85854-C4-3-R and by the EU H2020 EMPATHIC project grant number 769872.

References

1. Asher N, Lascarides A (2001) Indirect speech acts. Synthese 128(1):183–228. https://doi.org/10.1023/A:1010340508140
2. Chandramohan S, Geist M, Lefevre F, Pietquin O (2011) User simulation in dialogue systems using inverse reinforcement learning. Interspeech 2011:1025–1028
3. Cho K, van Merriënboer B, Bahdanau D, Bengio Y (2014) On the properties of neural machine translation: encoder–decoder approaches. Syntax Semant Struct Stat Transl, p 103
4. Core MG, Allen J (1997) Coding dialogs with the damsl annotation scheme. In: AAAI fall symposium on communicative action in humans and machines, Boston, MA, vol 56
5. Cuayáhuitl H, Renals S, Lemon O, Shimodaira H (2005) Human-computer dialogue simulation using hidden markov models. In: 2005 IEEE workshop on automatic speech recognition and understanding. IEEE, pp 290–295
6. Eckert W, Levin E, Pieraccini R (1997) User modeling for spoken dialogue system evaluation. In: Proceedings of the IEEE workshop on automatic speech recognition and understanding, 1997. IEEE, pp 80–87
7. Hancher M (1979) The classification of cooperative illocutionary acts. Lang Soc, pp 1–14
8. Henderson M, Thomson B, Williams J (2013) Dialog state tracking challenge 2 and 3 handbook. http://camdial.org/mh521/dstc
9. Hochreiter S, Schmidhuber J (1997) Long short-term memory. Neural Comput 9(8):1735–1780
10. Hurtado LF, Griol D, Sanchis E, Segarra E (2007) A statistical user simulation technique for the improvement of a spoken dialog system. Springer, Berlin, pp 743–752. https://doi.org/10.1007/978-3-540-76725-1_77
11. Kingma D, Ba J (2015) Adam: a method for stochastic optimization. In: Proceedings of the 3rd international conference on learning representations (ICLR), pp 1–13

12. Layla EA, Jing H, Suleman K (2016) A sequence-to-sequence model for user simulation in spoken dialogue systems. In: Interspeech
13. Levin E, Pieraccini R, Eckert W (2000) A stochastic model of human-machine interaction for learning dialog strategies. IEEE Trans Speech Audio Process 8(1):11–23
14. Pietquin O (2005) A framework for unsupervised learning of dialogue strategies. Presses univ. de Louvain
15. Pietquin O, Dutoit T (2006) A probabilistic framework for dialog simulation and optimal strategy learning. IEEE Trans Audio Speech Lang Process 14(2):589–599
16. Quarteroni S, González M, Riccardi G, Varges S (2010) Combining user intention and error modeling for statistical dialog simulators. In: INTERSPEECH, pp 3022–3025
17. Rieser V, Lemon O (2006) Cluster-based user simulations for learning dialogue strategies. In: Interspeech
18. Schatzmann J, Thomson B, Weilhammer K, Ye H, Young S (2007) Agenda-based user simulation for bootstrapping a pomdp dialogue system. In: Human language technologies 2007: the conference of the north american chapter of the association for computational linguistics; companion volume, Short Papers. Association for Computational Linguistics, pp 149–152
19. Schatzmann J, Weilhammer K, Stuttle M, Young S (2006) A survey of statistical user simulation techniques for reinforcement-learning of dialogue management strategies. Knowl Eng Rev 21(2):97–126
20. Scheffler K, Young S (2000) Probabilistic simulation of human-machine dialogues. In: Proceedings of the 2000 IEEE international conference on acoustics, speech, and signal processing, 2000, ICASSP'00, vol 2. IEEE, pp II1217–II1220

Yeah, Right, Uh-Huh: A Deep Learning Backchannel Predictor

Robin Ruede, Markus Müller, Sebastian Stüker and Alex Waibel

Abstract Using supporting backchannel (BC) cues can make human-computer interaction more social. BCs provide a feedback from the listener to the speaker indicating to the speaker that he is still listened to. BCs can be expressed in different ways, depending on the modality of the interaction, for example as gestures or acoustic cues. In this work, we only considered acoustic cues. We are proposing an approach towards detecting BC opportunities based on acoustic input features like power and pitch. While other works in the field rely on the use of a hand-written rule set or specialized features, we made use of artificial neural networks. They are capable of deriving higher order features from input features themselves. In our setup, we first used a fully connected feed-forward network to establish an updated baseline in comparison to our previously proposed setup. We also extended this setup by the use of Long Short-Term Memory (LSTM) networks which have shown to outperform feed-forward based setups on various tasks. Our best system achieved an F1-Score of 0.37 using power and pitch features. Adding linguistic information using word2vec, the score increased to 0.39.

Keywords Backchannels · Building rapport · Artificial intelligence
Speech recognition

1 Introduction

With dialog speech technology increasingly entering the mainstream of our every day lives (Siri, Cortana, Alexa, …), there is a growing interest in dialog systems that are not only utilitarian (to answer questions or carry out tasks), but also to entertain

R. Ruede · M. Müller (✉) · S. Stüker · A. Waibel
Karlsruhe Institute of Technology, Karlsruhe, Germany
e-mail: m.mueller@kit.edu

R. Ruede
e-mail: robin.ruede@student.kit.edu

A. Waibel
Carnegie Mellon University, Pittsburgh, PA, USA

M. Eskenazi et al. (eds.), *Advanced Social Interaction with Agents*, Lecture Notes in Electrical Engineering 510, https://doi.org/10.1007/978-3-319-92108-2_25

and to be social. Humanoid robots, interactive toys, virtual assistants and even virtual psychiatrists and pets attempt to add an emotional and social dimension to human interaction that may go beyond improving the user experience of existing dialog systems, and thus require increasingly skillful and adept social interaction. Social dialogs are, however, much less well understood than goal directed ones. They do not aim for a particular outcome other than the more indirect goals of growing a mutual understanding, empathizing, bonding and entertaining between humans.

In the present paper, we are proposing a neural network based system to generate a social response. Our first attempt in this regard aims to predict a suitable social response, when human speakers take "the floor" and are sharing thoughts and experiences. The so-called "backchannel" (BC) involves short phrases ("uh-huh", "hum", "yeah", right, etc.) whose role is to signal to another speaker that one is listening and paying attention. Further extensions also empathize, confirm, approve or disapprove. In conversational speech, BCs complement turn taking where more rapid questions and responses are exchanged. Despite its simple function, however, the BC is surprisingly complex: It must be chosen properly, timed correctly and placed at appropriate intervals. It also responds to content, emotion and discourse state.

In this paper, we describe a neural network approach to learning the production of proper BC cues. We will focus on short phrasal BC cues during longer stretches of conversational speech, where another speaker has taken the floor. Appropriate prediction of backchanneling is learned from human conversation and includes acoustic and linguistic features. In our work, we use recurrent neural networks to learn the choice and placement of appropriate BC cues from conversational data (Switchboard). Special attention is given to producing causal backchanneling, i.e., so that the generation of a BC can be produced in real-time systems with information of the past.

This paper is organized as follows: In the next Section, we provide an overview of related work. In Sect. 3, we describe our approach in detail, followed by an overview of the experimental setup in Sect. 4. The results of the experiments are presented in Sect. 5. This paper concludes with an outlook to future work in Sect. 6.

2 Related Work

Different approaches towards BC prediction have been proposed in the past. They are based on different types of predictors and use a wide variety of input modalities. These modalities include acoustic features like pause and pitch, but also visual cues like head movement. In addition to these direct features, additional information sources like language models or part of speech tagging exist.

Many approaches are rule based. Truong et al. [25] proposed a method that uses acoustic features. The authors state that the most important acoustic phenomena for BC prediction occur right before a BC. As features, they used pause information, as well as pitch (falling or rising slope). They conducted their experiments on a Dutch corpus and report that the most important feature in their work is the duration of the pause. Ward and Tsukahara [27] proposed a similar approach triggering BCs at low pitch and pause regions in English and Japanese. But building a rule-based system might prove difficult as these rules have to be manually created, which is time-consuming and difficult to generalize. Other works included data-driven methods in which a classifier is trained and the output of this classifier is then post-processed. Morency et al. [17] proposes an approach that incorporates sequential probabilistic models like Hidden Markov Models or Conditional Random Fields. They used a set of features including eye gaze and several features derived from the audio signal, e.g., downslopes in pitch or certain types of volume changes. In another approach, predicting different types of BC was attempted [8]. Detecting BCs in real-time was also proposed [20] in the past.

There exists another category of systems that make use of artificial neural networks (ANNs). Being a data-driven method, NNs do not require handwritten rules. They have shown to be a versatile tool with the ability to learn relevant features automatically. A first approach towards detecting speech acts (including BCs) was proposed by Ries [19]. He used an NN in combination with an HMM. Stolcke also proposed NN based methods for modelling dialogue acts [22, 23]. In the past, we also proposed an NN based approach [15] that was mainly data-driven, requiring only minimal post-processing of the network outputs. In this first approach, we used a very basic ANN based setup, which we now refined.

The objective evaluation of systems for BC prediction is difficult because BC behaviour is very speaker-dependent and subjective. As an objective measurement, the use of the F1-Score has been established. Kok and Heylen [11] provides a comparison of different approaches for evaluation. In addition to objective measures, user studies are also a possibility to evaluate BC systems, like we did in the past [15]. A general study about the occurrence of BCs with respect to their role in facilitating attentive listening also exists [7].

3 Backchannel Prediction

3.1 BC Utterance Selection

There are different kinds of phrasal BCs, they can be non-committal, positive, negative, questioning, et cetera. To simplify the problem of predicting BCs, we only try to predict the trigger times for any type of BC, ignoring the distinction between different kinds of responses.

3.2 Feature Selection

A neural network is able to learn advantageous feature representations on its own. Hence, feeding the absolute pitch and power (signal energy) values for a given time context enables the network to automatically extract the relevant information such as pitch slopes and pause triggers, as used in related research [17]. In addition to pitch and power, we also evaluated using other acoustic features such as the fundamental frequency variation (FFV) [12] and the Mel-frequency cepstral coefficients (MFCCs). Finally, we tried adding an encoding of the speakers' word history before the listener backchannel using word2vec [14] to assess whether our setup benefits from multimodal input features.

3.3 Training and Neural Network Design

We assumed to have two separate, but synchronized audio channels and corresponding transcripts: One for the speaker and one for the listener. We needed to decide which areas of audio to use to train the network. As we wanted to predict the BCs in an online fashion without using future information, we needed to train the network to detect segments of audio from the speaker track that would potentially cause a BC in the listener track. We chose the beginning of the BC utterance as an anchor and used a fixed context before that as the positive prediction area. We also needed to choose negative examples, so the network would not be biased to always predict a BC. We did this by selecting the range a few seconds before each BC, because in that area the listener explicitly decided not to give a backchannel response yet. This resulted in a fully balanced training dataset.

We initially used a feed forward network architecture. The input layer consists of all the chosen features over the previously selected fixed time context. The output layer has two softmax neurons representing the "categories" [BC, non-BC]. We used back-propagation to train the network on the outputs [1, 0] for BC and [0, 1] for non-BC prediction areas. We only need to consider one of these outputs because the softmax function guarantees that they add up to one. We evaluated multiple different combinations of network depths and neuron counts. An example of the architecture with two hidden layers can be seen in Fig. 1.

The placement of future BCs is dependent on the timing of previous BCs. The probability of a BC increases with longer periods without any listener feedback. To accommodate for this, we want the network to also take its previous internal state or outputs into account. We do this by modifying the above architecture to use Long-short term memory (LSTM) layers instead of feed forward layers.

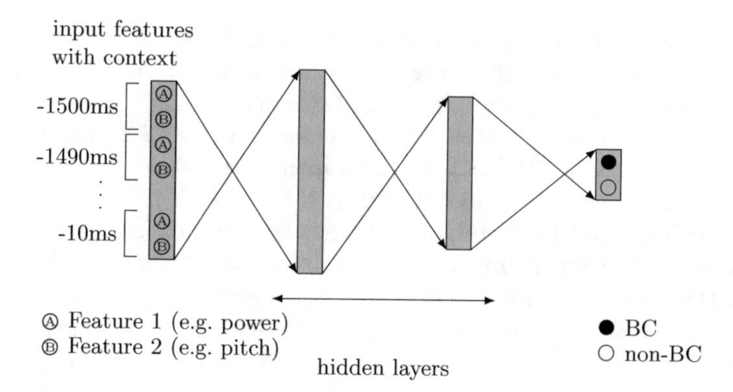

Fig. 1 Example for a neural network architecture for BC prediction

4 Experimental Setup

4.1 Dataset

We used the Switchboard dataset [3], which consists of 2,438 English telephone conversations of five to ten minutes, 260 h in total. Pairs of participants from across the United States were encouraged to talk about a specific topic chosen randomly from 70 possibilities. Conversation partners and topics were selected so two people would only talk once with each other, and every person would only discuss a specific topic once.

These telephone conversations are annotated with transcriptions and word alignments [4] with a total of 390 k utterances or 3.3 million words. We split the dataset randomly into 2,000 conversations for training, 200 for validation and 238 for evaluation. We used annotations from the Switchboard Dialog Act Corpus (SwDA) [6] to decide which utterances to classify as BCs. The SwDA contains categorical annotations for the utterances of about half of the data of the Switchboard corpus.

4.2 Extraction

We chose to use the top 150 most common unique utterances marked as BCs from the SwDA. Because the SwDA is incomplete, we had to identify utterances as BCs just by their text. We manually included some additional utterances that were missing from the SwDA transcriptions but present in the original transcriptions, by going through the most common utterances and manually selecting those that seemed relevant, such as 'um-hum yeah' and 'absolutely'. The most common BCs in the data set are "yeah", "um-hum", "uh-huh" and "right", adding up to 68% of all extracted BC phrases.

To select which utterances should be categorized as BCs and used for training, we first filtered noise and other markers such as laughter from the transcriptions. Some utterances such as "uh" can be both BCs and speech disfluencies, so we only chose those that have either silence or another BC before them. With this method a total of 15.7% of utterances or 2.21% of words were labelled as BCs.

We used the Janus Recognition Toolkit [13] for parts of the feature extraction (power, pitch tracking, FFV, MFCC). Features were extracted for 32 ms frame windows with a frame shift of 10 ms, resulting in 100 samples per feature dimension per second. Because most of the data does not change much every 10 ms, we also test different context strides by only extracting every n-th frame. As an example, 800 ms of context with a stride of 2 corresponds to 40 data frames. For word2vec, we chose to also emit one frame every 10 ms for consistency, containing the encoding of the last non-silent word that ended before or at the time of the frame.

4.3 Training

We used Theano [24] with Lasagne [1] for rapid prototyping and testing of different parameters.[1] To evaluate different hyperparameters, we trained multiple network configurations with various context lengths (500–2000 ms), context strides (1–4 frames), network depths (one to four hidden layers), layer sizes (15–125 neurons), activation functions (tanh and relu), optimization methods (SGD, Adadelta and Adam [9]), weight initialization methods (constant zero and Glorot [2]), and layer types (feed forward and LSTM).

The LSTM networks we tested were prone to overfitting quickly. We tried two methods of regularization to overcome this. The first was Dropout training, where we randomly dropped a specific portion of neuron outputs in each layer for each training batch [21]. We evaluated dropout layer combinations from 0 to 50% while increasing layer sizes proportionately, but this did not improve the results. The second was adding L2-Regularization with a constant factor of 0.0001. This greatly reduced overfitting and slightly improved the results.

4.4 Postprocessing

We interpret the output value of the neural networks as the probability of a BC occurring at a given time. As the output is very noisy, we first apply a low-pass filter. To ensure our prediction does not use any future information, we use a Gaussian filter which is asymmetrically cut off at some multiple c of the standard deviation σ

[1]Our code for extraction, training, postprocessing and evaluation is available at https://github.com/phiresky/backchannel-prediction. The repository also contains a script to reproduce all of the results of this paper.

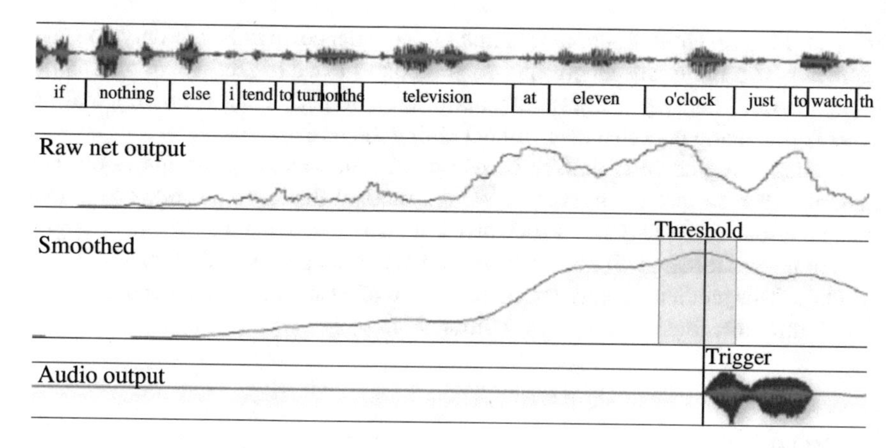

Fig. 2 Postprocessing example

for the side that would range into the future, and offset it so the last frame is at ±0 ms from the prediction target time. This means the latency of our prediction increases by $c \cdot \sigma$ ms. If we choose $c = 0$, we cut off the complete right half of the bell curve, keeping the latency at 0 at the cost of accuracy of the filter.

After the low-pass filter, we select every area for which the value exceeds a given threshold. We trigger either at the beginning of each of these areas or at their first local maximum, depending on the largest acceptable latency. This varies depending on the chosen allowed margin of error as defined in Sect. 4.5. An example of this postprocessing process can be seen in Fig. 2.

We determined the optimal postprocessing hyperparameters for each network configuration and allowed margin of error automatically using Bayesian optimization [16] with the validation F1-Score as the utility function. For a margin of error of [0 ms, +1000 ms], the resulting standard deviation σ ranged from 200 to 350 ms, and the filter cut-off ranged from 0.9σ to 1.4σ. With this margin of error, the prediction can happen with a delay of up to one second after the ground truth. When choosing a margin of error that only allows a smaller delay such as [−200 ms, +200 ms], the selected standard deviation ranged from 150 to 250 ms, and the filter cut-off ranged from 0.0σ to 1.0σ, causing the predicted trigger to happen earlier.

4.5 Evaluation

The training data contains two-sided conversations. Because the output of our predictor is only relevant for segments with just one person talking, we run our evaluation on monologuing segments. We define a monologuing segment as the maximum possible time range in which one person is continuously *talking* and the other person is continuously *not talking* for at least five seconds. A person is continuously talking *iff* they are only emitting utterances that are not silence or BCs. We only consider segments of a minimum length of five seconds to exclude sections of alternating conversation. This way we get segments where one participant is talking, and the other is producing backchannels without taking the turn.

We define a prediction time as correct if it is within a given margin of error of the onset of a correct BC utterance. We did most of the testing of our predictions with an error margin of [0 ms, +1000 ms], but also provide results for other margins used in related research. For comparison, we also evaluated a random predictor as a baseline. This predictor knows the correct count of BCs for an audio file and returns a uniformly distributed set of trigger times.

5 Results

We use "70 : 35" to denote a network layer configuration of input → 70 neurons → 35 neurons → output.

We tested different context widths. A context width of n ms means we use the range [$-n$ ms, 0 ms] from the beginning of the backchannel utterance. The results improved when increasing the context width from our initial value of 500 ms. Performance peaked with a context of about 1500 ms, as can be seen in Table 1a, longer contexts tended to cause the predictor to trigger too late. We tested using only every n-th frame of input data. Even though we initially did this for performance reasons, we noticed that training on every single frame has worse performance than skipping every second frame due to overfitting. Taking every fourth frame seems to miss too much information, so performance peaks at a context stride of 2, as seen in Table 1b.

We tested different combinations of features, using solely power in a first approach. But adding prosodic features gives great improvement. Using FFV as the only prosodic feature performs worse than FFV together with the absolute pitch value. Adding MFCCs does not seem to improve performance in a meaningful way, when also using pitch, see Table 1c for more details. Note that using *only* word2vec performs reasonably well, because with our method it indirectly encodes the time since the last utterance, similar to the power feature. Table 1d shows a comparison between feed forward and LSTM networks. The parameter count is the number of connection weights the network learns during training. Note that LSTMs have higher performance, even with similar parameter counts. We compared different layer sizes for our LSTM networks, as shown in Table 1e. A network depth of two hidden layers worked best, but the results are adequate with a single hidden layer or three hidden layers.

In Table 2, our final results are given for the completely independent evaluation data set. We compared the results from [15] with our system. Müller et al. [15] used the same dataset, but focused on offline predictions, meaning their network had future information available, and they evaluated their performance on the whole corpus including segments with silence and with alternating conversation. We adjusted

Table 1 Results on the validation set. All results use the following setup if not otherwise stated: LSTM, configuration: (70 : 35); input features: power, pitch, ffv; context width: 1500 ms; context frame stride: 2; margin of error: 0 ms to +1000 ms. Precision, recall, and F1-Score are given for the validation data set.

(a) Results with various context lengths. Performance peaks at 1500 ms

Context	Precision	Recall	F1-Score
500 ms	0.219	0.466	0.298
1000 ms	0.280	0.497	0.358
1500 ms	0.305	0.488	**0.375**
2000 ms	0.275	0.577	0.373

(b) Results with various context frame strides

Stride	Precision	Recall	F1-Score
1	0.290	0.490	0.364
2	0.305	0.488	**0.375**
4	0.285	0.498	0.363

(c) Results with various input features, separated into only acoustic features and acoustic plus linguistic features

Features	Precision	Recall	F1-Score
power	0.244	0.516	0.331
power, pitch	0.307	0.435	0.360
power, pitch, mfcc	0.278	0.514	0.360
power, ffv	0.259	0.513	0.344
power, ffv, mfcc	0.279	0.515	0.362
power, pitch, ffv	0.305	0.488	**0.375**
word2vec$_{dim=30}$	0.244	0.478	0.323
power, pitch, word2vec$_{dim=30}$	0.318	0.486	0.385
power, pitch, ffv, word2vec$_{dim=15}$	0.321	0.475	0.383
power, pitch, ffv, word2vec$_{dim=30}$	0.322	0.497	**0.390**
power, pitch, ffv, word2vec$_{dim=50}$	0.304	0.527	0.385

(d) Feed forward versus LSTM. LSTM outperforms feed forward architectures

Layers	Parameter count	Precision	Recall	F1-Score
FF (56 : 28)	40k	0.230	0.549	0.325
FF (70 : 35)	50k	0.251	0.468	0.327
FF (100 : 50)	72k	0.242	0.490	0.324
LSTM (70 : 35)	38k	0.305	0.488	**0.375**

(e) Comparison of different network configurations. Two LSTM layers give the best results

Layer sizes	Precision	Recall	F1-Score
100	0.280	0.542	0.369
50 : 20	0.291	0.506	0.370
70 : 35	0.305	0.488	**0.375**
100 : 50	0.303	0.473	0.369
70 : 50 : 35	0.278	0.541	0.367

Table 2 Final best results on the evaluation set (chosen by validation set)

(a) Comparison with previous research. Müller et al. [15] did their evaluation without the constraints defined in Sect. 4.5, so we adjusted our baseline and evaluation to match their setup

Predictor	Precision	Recall	F1-Score
Baseline (random)	0.042	0.042	0.042
Müller et al. (offline) [15]	–	–	0.109
Our results (offline, context of −750–750 ms)	0.114	0.300	**0.165**
Our results (online, context of −1500–0 ms)	0.100	0.318	0.153

(b) Results with our evaluation method with various margins of error used in other research [10]. Performance improves with a wider margin width and with a later margin center

Margin of Error	Constraint	Precision	Recall	F1-Score
−200 to +200 ms		0.172	0.377	0.237
−100 to +500 ms		0.239	0.406	0.301
−500 to +500 ms		0.247	0.536	0.339
0 to +1000 ms	Baseline (random, correct BC count)	0.111	0.052	0.071
	Baseline (random, 8x correct BC count)	0.079	0.323	0.127
	Balanced Precision and Recall	0.342	0.339	0.341
	Best F1-Score (only acoustic features)	0.294	0.488	0.367
	Best F1-Score (acoustic and linguistic features)	0.312	0.511	**0.388**

our baseline and evaluation system to match their setup by removing the monologuing constraint described in Sect. 4.5 and changing the margin of error to [−200 ms, +200 ms]. As can be seen in Table 2a, our predictor performs better. All other related research used different languages, datasets or evaluation methods, rendering a direct comparison difficult because of slightly different tasks.

Table 2b shows the results with our presented evaluation method. We provide scores for different margins of error used in other research. Subjectively, missing a BC trigger may be more acceptable than a false positive, so we also provide a result with balanced precision and recall. Note that a later margin center with the same margin width has higher performance because it allows more latency in the predictions, which means we can choose better postprocessing parameters as described in Sect. 4.4.

6 Conclusion and Future Work

We have presented a new approach to predict BCs using neural networks. With refined methods for network training as well as different network architectures, we could improve the F1-Score in contrast to our previous experiments. In addition to evaluating different hyperparameter configurations, we also experimented with LSTM networks, which lead to improved results. Our best system achieved an F1-Score of 0.388.

We used linguistic features via word2vec only in a very basic way, assuming the availability of an instant speech recognizer by using the reference transcripts. As low-latency speech recognition is possible [18], one of the next steps would be to combine both systems. Further work is needed to evaluate other methods for adding word vectors to the input features and to analyze problems with our approach. We only tested feed forward neural networks and LSTMs, other architectures like time-delay neural networks [26], also called convolutional neural networks, may also give interesting results. Our approach of choosing the training areas may not be optimal because the delay between the last utterance of the speaker and the backchannel can vary significantly. One possible solution would be to align the training area by the last speaker utterance instead of the backchannel start.

Because backchannels are a largely subjective phenomenon, a user study would be helpful to subjectively evaluate the performance of our predictor and to compare it with human performance in our chosen evaluation method. Another method would be to find consensus for backchannel triggers by combining the predictions of multiple human subjects for a single speaker channel as described by Huang et al. (2010) as "parasocial consensus sampling" [5].

Acknowledgements This work has been conducted in the SecondHands project which has received funding from the European Unions Horizon 2020 Research and Innovation programme (call:H2020-ICT-2014-1, RIA) under grant agreement No 643950.

References

1. Dieleman S, Schlter J, Raffel C, Olson E, Sønderby SK et al (2015) Lasagne: first release. https://doi.org/10.5281/zenodo.27878
2. Glorot X, Bengio Y (2010) Understanding the difficulty of training deep feedforward neural networks. Aistats 9:249–256
3. Godfrey J, Holliman E (1993) Switchboard-1 release 2. https://catalog.ldc.upenn.edu/ldc97s62
4. Harkins D et al (2003) ISIP switchboard word alignments. https://www.isip.piconepress.com/projects/switchboard/
5. Huang L, Morency LP, Gratch J (2010) Learning backchannel prediction model from parasocial consensus sampling: a subjective evaluation. In: International conference on intelligent virtual agents. Springer, pp 159–172
6. Jurafsky D, Van Ess-Dykema C et al (1997) Switchboard discourse language modeling project
7. Kawahara T, Uesato M, Yoshino K, Takanashi K (2015) Toward adaptive generation of backchannels for attentive listening agents. In: International workshop serien on spoken dialogue systems technology, pp 1–10

8. Kawahara T, Yamaguchi T, Inoue K, Takanashi K, Ward N (2016) Prediction and generation of backchannel form for attentive listening systems. In: Proceedings of the INTERSPEECH, vol 2016
9. Kingma D, Ba J (2014) Adam: a method for stochastic optimization. arXiv preprint arXiv:1412.6980
10. de Kok I, Heylen D (2012) A survey on evaluation metrics for backchannel prediction models. In: Proceedings of the interdisciplinary workshop on feedback behaviors in dialog
11. Kok ID, Heylen D (2012) A survey on evaluation metrics for backchannel prediction models. In: Feedback behaviors in dialog
12. Laskowski K, Heldner M, Edlund J (2008) The fundamental frequency variation spectrum. Proc Fon 2008:29–32
13. Levin L, Lavie A, Woszczyna M, Gates D, Gavaldá M, Koll D, Waibel A (2000) The janus-iii translation system: speech-to-speech translation in multiple domains. Mach Trans 15(1):3–25. https://doi.org/10.1023/A:1011186420821
14. Mikolov T, Chen K, Corrado G, Dean J (2013) Efficient estimation of word representations in vector space. arXiv preprint arXiv:1301.3781
15. Müller M, Leuschner D, Briem L, Schmidt M, Kilgour K, Stüker S, Waibel A (2015) Using neural networks for data-driven backchannel prediction: a survey on input features and training techniques. In: International conference on human-computer interaction. Springer, pp 329–340
16. Mockus J (1974) On bayesian methods for seeking the extremum. In: Proceedings of the IFIP technical conference. Springer, London, pp 400–404. http://dl.acm.org/citation.cfm?id=646296.687872
17. Morency LP, de Kok I, Gratch J (2010) A probabilistic multimodal approach for predicting listener backchannels. Auton Agent Multi-Agent Syst 20(1):70–84. https://doi.org/10.1007/s10458-009-9092-y
18. Niehues J, Nguyen TS, Cho E, Ha TL, Kilgour K, Müller M, Sperber M, Stüker S, Waibel A (2016) Dynamic transcription for low-latency speech translation. Interspeech 2016:2513–2517
19. Ries K (1999) HMM and neural network based speech act detection. In: Proceedings of the IEEE international conference on acoustics, speech, and signal processing, 1999, vol 1. IEEE Computer Society, pp 497–500
20. Schroder M, Bevacqua E, Cowie R, Eyben F, Gunes H, Heylen D, Ter Maat M, McKeown G, Pammi S, Pantic M et al (2012) Building autonomous sensitive artificial listeners. IEEE Trans Affect Comput 3(2):165–183
21. Srivastava N, Hinton GE, Krizhevsky A, Sutskever I, Salakhutdinov R (2014) Dropout: a simple way to prevent neural networks from overfitting. J Mach Learn Res 15(1):1929–1958
22. Stolcke A, Ries K, Coccaro N, Shriberg E, Bates R, Jurafsky D, Taylor P, Martin R, Van Ess-Dykema C, Meteer M (2000) Dialogue act modeling for automatic tagging and recognition of conversational speech. Comput Linguist 26(3):339–373
23. Stolcke A, et al (1998) Dialog act modeling for conversational speech. In: AAAI spring symposium on applying machine learning to discourse processing, pp 98–105
24. Theano Development Team: Theano: a python framework for fast computation of mathematical expressions (2016). arXiv e-prints http://arxiv.org/abs/1605.02688
25. Truong KP, Poppe RW, Heylen DKJ (2010) A rule-based backchannel prediction model using pitch and pause information. In: Proceedings of the interspeech 2010, Makuhari, Chiba, Japan. International Speech Communication Association (ISCA), pp 3058–3061
26. Waibel A, Hanazawa T, Hinton G, Shikano K, Lang KJ (1989) Phoneme recognition using time-delay neural networks. IEEE Trans Acoust Speech Signal Process 37(3):328–339
27. Ward N, Tsukahara W (2000) Prosodic features which cue back-channel responses in English and Japanese. J Pragmat 32(8):1177–1207

Printed in the United States
By Bookmasters